Mechanical Stress Evaluation by Neutron and Synchrotron Radiation
MECA SENS 2017

Edited by

Deon Marais
Thomas M. Holden
Andrew M. Venter

This book presents the proceedings of the 9th International Conference on Mechanical Stress Evaluation by Neutron and Synchrotron Radiation which was hosted by the South African Nuclear Energy Corporation (Necsa) SOC Limited in cooperation with the International Atomic Energy Agency (IAEA). The conference topics included deformation & modeling, processing & welding, techniques & instruments, mechanical methods vs. diffraction, microstructure & characterization, surface modification & coatings, 3D/4D characterization and fatigue, creep & plasticity.

In dedication to Philip Doubell [†]

Mechanical Stress Evaluation by Neutron and Synchrotron Radiation
MECA SENS 2017

Proceedings of the 9th International Conference on Mechanical Stress Evaluation by Neutron and Synchrotron Radiation

Kruger National Park, South Africa
19-21 September 2017

**Deon Marais, Thomas M. Holden
and Andrew M. Venter**

Research and Development Division, Necsa SOC Limited
South African Nuclear Energy Corporation, PO Box 582
Pretoria 0001, South Africa

Peer review statement

All papers published in this volume of "Materials Research Proceedings" have been peer reviewed. The process of peer review was initiated and overseen by the above proceedings editors. All reviews were conducted by expert referees in accordance to Materials Research Forum LLC high standards.

Published under License by **Materials Research Forum LLC**
Millersville, PA 17551, USA

Published as part of the proceedings series
Materials Research Proceedings
Volume 4 (2018)

ISSN 2474-3941 (Print)
ISSN 2474-395X (Online)

ISBN 978-1-94529166-1 (Print)
ISBN 978-1-94529167-8 (eBook)

This book contains information obtained from authentic and highly regarded sources. Reasonable efforts have been made to publish reliable data and information, but the author and publisher cannot assume responsibility for the validity of all materials or the consequences of their use. The authors and publishers have attempted to trace the copyright holders of all material reproduced in this publication and apologize to copyright holders if permission to publish in this form has not been obtained. If any copyright material has not been acknowledged please write and let us know so we may rectify in any future reprint.

Distributed worldwide by

Materials Research Forum LLC
105 Springdale Lane
Millersville, PA 17551
USA
http://www.mrforum.com

Manufactured in the United State of America
10 9 8 7 6 5 4 3 2 1

Table of Contents

Processing & Welding

Surface Modification and Coating

Techniques & Instruments

Others

Preface

The 9th International Conference on Mechanical Stress Evaluation by Neutron and Synchrotron Radiation (MECA SENS 2017), was held at the Skukuza Rest Camp of the Kruger National Park, South Africa over the period 19th to 21st of September 2017.

MECA SENS 2017 was the latest of a highly successful series which started in Reims (France) in 2000 and continued in Manchester (UK, 2003), Santa Fe (USA, 2005), Vienna (Austria, 2007), Mito (Japan, 2009), Hamburg (Germany, 2011), Sydney (Australia, 2013) and Grenoble (France, 2015).

The conference offered a unique opportunity to overview the most recent developments and capabilities of diffraction based approaches, complemented by mechanical and image-based techniques as applicable to residual stress determination. Participants were scientists and engineers from academia, research facilities and industry spanning the scope from well experienced practitioners, instrumentation experts and innovators, to postgraduate students and researchers new to this exciting field.

Feedback received from attendees indicated that MECA SENS 2017 was a highly successful conference with respect to not only the inimitable social activities, but also the exceptional scientific standard of the lectures. There were 68 oral and 23 poster presentations delivered to 95 delegates from 18 countries and 52 institutions.

Publication in these proceedings was voluntary and after a peer review process, a total of 25 papers were accepted for publication. For this I would like to thank the authors and reviewers on behalf of the MECA SENS community.

Finally, I would like to thank the members of the MECA SENS International Scientific Committee, the MECA SENS 2017 Program Advisory Committee and the Local Organizing Committee without whose tireless work MECA SENS 2017 would not have been the success that it was.

Andrew Venter
Chair, MECA SENS 2017

March, 2018
Pretoria, South Africa.

Committees

Local Organizing Committee:
- Andrew Venter: *Chairman* (Research & Development, Necsa SOC Ltd)
- Mihloti Baloyi: *Secretary* (Research & Development, Necsa SOC Ltd)
- Tshepo Ntsoane (Research & Development, Necsa SOC Ltd)
- Deon Marais (Research & Development, Necsa SOC Ltd)
- Sheryl van den Berg: *Conference organizer* (The Inside Edge)

Programme Advisory Committee:
- Anthonie Cilliers (School of Mechanical & Nuclear Engineering, North West University)
- Axel Steuwer (Nelson Mandela Metropolitan University, Port Elizabeth)
- Claudia Polese (School of Mechanical, Industrial and Aeronautical Engineering, University of the Witwatersrand)
- Daniel Francis (Pressure Vessels and Heat Exchangers, Sasol Group Technology)
- Daniel Hattingh (Nelson Mandela Metropolitan University, Port Elizabeth)
- Esther Akinlabi (Faculty of Engineering and the Built Environment, University of Johannesburg)
- Mark Newby (Sustainability Group, Eskom Holdings SOC Ltd.)
- Philip Doubel (Sustainability Group, Eskom Holdings SOC Ltd.)
- Pieter Pistorius (Department of Materials Science and Metallurgical Engineering, University of Pretoria)
- Sisa Pityana (National Laser Centre, Council for Scientific and Industrial Research)
- Thorsten Becker (Department of Mechanical and Mechatronic Engineering, University of Stellenbosch)

International Scientific Committee:
- Alain Lodini (University of Reims, France)
- Andreas Schreyer (European Spallation Source, Sweden)
- Andrew Venter (Necsa SOC Ltd, South Africa)
- Axel Steuwer (University of Malta, Spain)

- Chedly Braham (PIMM, France)
- Cevdet Noyan (Columbia University, USA)
- Wan Chuck Woo (KAERI, South Korea)
- Daigo Setoyama (Toyota Central Research and Development Laboratories, Japan)
- Donald Brown (LANL, USA)
- Hahn Choo (University of Tennessee, USA)
- Hiroshi Suzuki (JAEA, Japan)
- Jens Gibmeier (KIT, Germany)
- Jette Oddershede (FYSIK, Denmark)
- Jon Almer (APS, USA)
- Ke An (ORNL, USA)
- Keisuke Tanaka (Meiji University, Japan)
- Krzystof Wierzbanowski (AGH University of Science and Technology, Poland)
- Lyndon Edwards (ANSTO, Australia)
- Mark Bourke (LANL, USA)
- Mark Daymond (Queen's University, Canada)
- Michael Fitzpatrick (Coventry University, UK)
- Ondrej Muransky (ANSTO, Australia)
- Petr Lukas (UJF, Czech Republic)
- Phil Withers (University of Manchester, UK)
- Ron Rogge (CNL, Canada)
- Shu Yan Zhang (ISIS, UK)
- Thilo Pirling (ILL, France)
- Thomas Buslaps (ESRF, France)
- Walter Reimers (TU-Berlin, Germany)
- Xun-Li Wang (The City University of Hong Kong, China)
- Yo Tomota (Ibaraki University, Japan)
- Yoshiaki Akiniwa (Yokohama National University, Japan)

Sponsors

Necsa SOC Limited
Elias Motsoaledi Street Extension West,
North-West Province, South Africa
www.necsa.co.za

International Atomic Energy Agency
Vienna International Centre, Austria
www.iaea.org

SA/UK Newton Fund
www.nrf.ac.za/tags/newton-fund

National Research Foundation
CSIR Complex, Meiring Naudé Road,
Brummeria, Pretoria, South Africa
www.nrf.ac.za

ISIS Neutron and Muon Source
Science and Technology Facilities Council,
Didcot, United Kingdom
www.isis.stfc.ac.uk

**DST-NRF Centre of Excellence in Strong
Materials**
University of the Witwatersrand,
Braamfontein, Johannesburg, South Africa
www.wits.ac.za/strongmaterials

DECTRIS Limited
Taefernweg 1, 5405 Baden-Daettwil,
Switzerland
www.dectris.com

Keynote

MECA SENS 2017　　　　　　　　　　　　　　　　Materials Research Forum LLC
Materials Research Proceedings **4** (2018) 3-8　　　doi: http://dx.doi.org/10.21741/9781945291678-1

Diffraction Methods and Scale Transition Model used to study Evolution of Intergranular Stress and Micro-Damage Phenomenon during Elasto-Plastic Deformation

A. Baczmański[1,a*], S. Wroński[1,b], E. Gadalińska[2,c], Y. Zhao[3,d], L. Le Joncour[3,e], C. Braham[4,f], C. Scheffzük[6,7,g] and P. Kot[1,h]

[1]AGH-University of Science and Technology, WFiIS, al. Mickiewicza 30, 30-059 Krakow, Poland

[2]Institute of Aviation, al. Krakowska 110/114, 02-256 Warszawa, Poland

[3]ICD-LASMIS, Université de Technologie de Troyes (UTT), 12 rue Marie Curie, 10004 Troyes, France

[4]PIMM, Arts et Métiers ParisTech (ENSAM), 151 Bd de l'Hôpital, 75013 Paris, France

[5]FLNP, Joint Institute for Nuclear Research, 141980 Dubna, Russia

[6]Karlsruhe Institute of Technology, Adenauerring 20b, 76131 Karlsruhe, Germany

[a]Baczmanski@fis.agh.edu.pl, [b]Wronski@fis.agh.edu.pl, [c]Elzbieta.Gadalinska@ilot.edu.pl, [d]Yuchen.Zhao@utt.fr, [e]Lea.le_Joncour@utt.fr, [f]Chedly.Braham@ensam.eu, [g]Christian.Scheffzuek@kit.edu, [h]Przemyslaw.Kot@fis.agh.edu.pl

Keywords: Polycrystalline Material, Yield Condition, Damage Process, Self-Consistent Model, Diffraction Measurements

Abstract. A methodology combining diffraction experiments and self-consistent calculations was used to study the mechanical behaviour of groups of grains within two-phase polycrystalline materials. In this work, an Al/SiC$_p$ composite and duplex austenitic-ferritic steel are studied. The lattice strain evolution was determined from lattice strain measured in situ during tensile tests using neutron diffraction. The experimental results were used to study slip on crystallographic planes, localisation of stresses in polycrystalline grains and the mechanical effects of damage occurring during plastic deformation. For this purpose, a prediction made using the recently developed new version of the elasto-plastic self-consistent model was compared with the experimental data.

Introduction

Diffraction methods for lattice strain measurement provide useful information concerning the nature of grains behaviour during elastoplastic deformation. The main advantage of the diffraction methods is the possibility of studying mechanical properties of polycrystalline materials separately in each phase and groups of grains with a specific orientation. These methods enable an analysis of macrostress and microstress for multiphase and anisotropic materials. The multi-scale crystallographic models are very convenient for the study of elasto-plastic properties on microscopic and macroscopic scales. Comparison of experimental data with model predictions allows us to understand the physical phenomena, which occur during sample deformation at the level of polycrystalline grains. Moreover, the micro and macro parameters of elasto-plastic deformation can be experimentally established. The main advantage of the methodology combining diffraction experiment and the self-consistent calculation is that the mechanical behaviour of polycrystalline groups of grains or different phases can be studied.

In this work, neutron diffraction was used to study *in situ* deformation of two phases in an Al/SiC$_p$ composite and duplex stainless steels during tensile loading. The aim is to show the role of reinforcement in the partitioning of loads between phases in metal matrix composites (MMC). Next, the partitioning of the stresses between two phases of elasto-plastically deformed duplex steel is studied, and attention is paid to stress relaxation indicating damage processes. Interpretation of experimental results is done using the self-consistent model including prediction of damage process.

Self-consistent model including damage prediction

The lattice strains measured *in situ* during the diffraction experiment can be compared with calculations performed using self-consistent models in which the homogenization method based on the interaction of an ellipsoidal inclusion with the homogenous medium is considered [1]. In many works the theoretical results were obtained through the self-consistent model of elastoplastic deformation based on formalism proposed by Hill [2] and developed by Turner & Tome [3]. This method was implemented for the interpretation of the diffraction experiment by Clausen *et al.* [4] and used in works such as [5, 6].

Another formulation of the self-consistent elastoplastic model was proposed by Lipinski & Berveiller [7]. Despite the differences in the constitutive equations and homogenization scheme, in both self-consistent elastoplastic models the interaction of an ellipsoidal inclusion with the homogenous medium is approximated by the Eshelby tensor [1]. The model developed by Berveiller and Lipinski describes the behaviour of a polycrystalline material for large strains, taking the rotation of the crystal lattice into account. The latter method was used by many authors to predict elastoplastic deformation and texture evolution in polycrystalline materials [8-11].

Recently, the self-consistent model (version by Lipinski & Berveiller) was developed to predict ductile micro-damage process [10]. To do this, the assumption of total energy equivalence [11] was applied at the grain scale and the effective total strain $\tilde{\varepsilon}_{ij}^g$ and the effective stress $\tilde{\sigma}_{ij}^g$ tensors were introduced for each grain g:

$$\tilde{\varepsilon}_{ij}^g = \varepsilon_{ij}\sqrt{1 - d^g} \text{ and } \tilde{\sigma}_{ij}^g = \frac{\sigma_{ij}^g}{\sqrt{1 - d^g}} \tag{1}$$

where: d^g is a scalar damage variable which describes damage at a grain scale.

Assuming that d^g in Eq. 1 depends on total strain and stress tensors, the expression for tangent moduli and strain concentration tensor were defined for the damaged material using comparison with the equivalent undamaged material [11, 12]. In this approach also the influence of damage on the evolution of critical resolved shear stress (CRSS, denoted by τ_c) and hardening parameter (H) are taken into account (for details see [11]). The physical consequences of the damage process occurring in a given grain are a decrease in localized stress and an increase in total deformation, which in turn will lead to softening of the grain.

Finally, to describe micro-damage process occurring in the grain g the variation of the d^g function has to be established, which is defined by $\dot{d}^g$ rate according to the following relation:

$$\dot{d}^g = \xi^{ph}\left(\varepsilon_{eq}^g - \varepsilon_0^{ph}\right)_+^{n^{ph}} \left(\dot{\varepsilon}_{eq}^g\right)_+ \tag{2}$$

where: ε_{eq}^g is the second invariant of the total strain tensor for a grain g and $\varepsilon_0^{ph}, n^{ph}, \xi^{ph}$ are phase-dependent parameters (denoted by the superscript ph), $(...)_+$ are the Macaulay brackets, which means the positive part of the quantity ε, i.e. $(x)_+ = x$ if $x > 0$ and $(x)_+ = 0$ if $x \le 0$.

MECA SENS 2017   Materials Research Forum LLC
Materials Research Proceedings **4** (2018) 3-8   doi: http://dx.doi.org/10.21741/9781945291678-1

Sources of hardening in Al/SiC$_p$ composite

The experiments analysed in this work were performed for the Al/SiC$_p$ metal matrix composite comprising 2124 aluminium alloy and ultrafine particles of silicon carbide (size of 0.7 μm). It was produced by a powder metallurgy route comprising a blending of the alloy powder and reinforcement, compaction, and consolidation by hot isostatic pressing. The amount of the reinforcement particles was 17.8 % by volume.

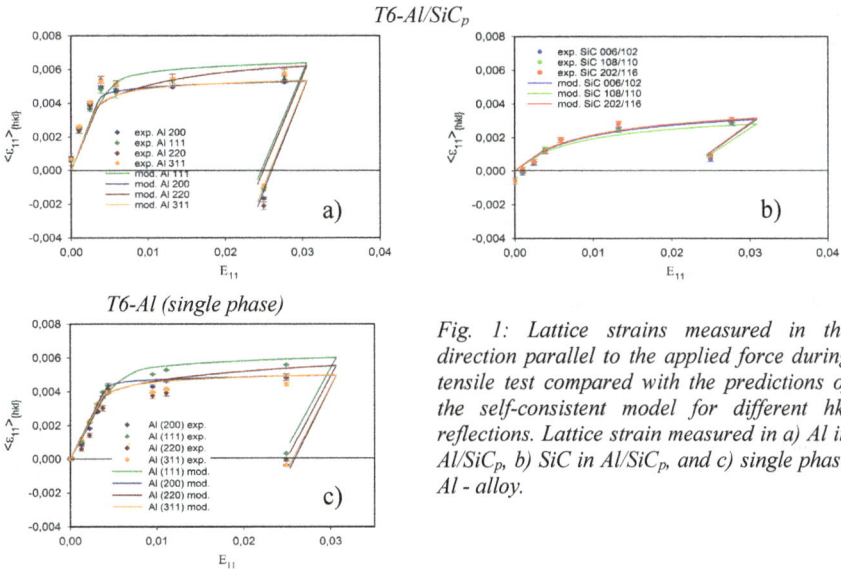

Fig. 1: Lattice strains measured in the direction parallel to the applied force during tensile test compared with the predictions of the self-consistent model for different hkl reflections. Lattice strain measured in a) Al in Al/SiC$_p$, b) SiC in Al/SiC$_p$, and c) single phase Al - alloy.

The composite specimens were examined after T6 heat treatment, i.e., it was solution treated at 491°C for 6 h and then water quenched and artificially aged for 4 h at 191°C. The specimen was subjected to *in situ* tensile tests. To perform comparative measurements a specimen of pure aluminium 2124 after T6 heat treatment was also prepared.

The *in situ* tensile test was performed with the time-of-flight (TOF) method on the EPSILON-MDS instrument in the JINR in Dubna (Russia). The lattice strains were gathered at the ambient temperature with two detector sets enabling measurements in two directions: in the direction of applied force and the perpendicular direction. The measurements were performed for 8 stages of deformation, as well as for the initial and residual state of the material (each point was measured during about 22 h, after stabilisation of the applied load). In Fig. 1 the results obtained in the direction of the applied load vs. sample strain are shown. The experimental stresses in the initial Al/SiC$_p$ sample (and corresponding lattice strains seen for zero load in Figs. 1a and 1b) were determined in both phases using 9 detectors at EPSILON-MDS instrument. As the reference the stress free lattice parameter measured for SiC powder was used. This can be done because the structure of SiC does not undergo phase transformation during production and thermal treatment of the composite. On the other hand the lattice parameter of Al powder cannot be taken as the reference due to precipitation processes occurring in the alloy during thermal treatment. Therefore, the value of hydrostatic stress for Al matrix (and corresponding stress free parameter of Al) was estimated from hydrostatic stress SiC, assuming equilibrium of stresses between both phases.

The self-consistent modelling results are also presented in Fig. 1 for both measured samples (strains in the perpendicular direction are not shown here, but they also agree with model prediction). The single crystal elastic constants of pure aluminium and H6 polytype of SiC [13] were used in calculations. In the case of Al/SiC$_p$ specimen, the agreement between experimental results and modelling was obtained for CRSS $\tau_{0Al} = 120$ MPa and hardening parameter $H_{Al} = 50$ MPa (SiC particles remained elastic during whole deformation). The same values of model parameters were obtained from a comparative experiment performed for aluminium alloy 2124 subjected to the same thermal treatment (T6) as Al/SiC$_p$ specimen.

On the basis of the lattice strain evolution in the Al/SiC$_p$ composite it can be stated that the partitioning of load between Al matrix and SiC reinforcement is well predicted for advanced plastic deformation and after samples unloading. At the beginning of the tensile test, including elastic range and elastic-plastic transition, the relaxation of initial inter-phase stresses occurs, and this process is not reproduced by the model used. This effect can be caused by micro-damage/decoupling process at the interfaces of SiC particles and Al matrix. Comparing the results for Al/SiC$_p$ and Al-alloy (singe phase) samples, nearly the same plastic behaviour was found for the aluminium phase (cf. Fig. 1a and 1c). It was also found that the evolution of lattice strains is similar for different hkl reflections in SiC phase (Fig. 1a), while significant difference between lattice strains in Al phase occurs during plastic deformation (Fig. 1b). The latter effect can be explained due to plastic anisotropy of Al grains and this is also seen in the case of single phase Al-alloy (Fig. 1c). The effects of anisotropy as well as the partitioning of the stresses between phases are well predicted by the model used.

Damage process in stainless duplex steel

The studied UR45N duplex steel is composed of ferrite and austenite, with the volume fraction of each phase approximately equal to 50 %. The steel was annealed at a temperature of 1050 °C and quenched with water to avoid precipitation of secondary phases. Finally, it was aged at 400°C for 1000 h and subsequently cooled in air at the ambient conditions.

Time of flight (TOF) neutron diffraction was used on the ENGIN-X instrument at the ISIS spallation neutron source to measure the lattice strains in the examined duplex steel. The size of the incident beam was limited by a slit (4 mm wide and 8 mm high), while the exit aperture of 4 mm was defined by radial collimators. The lattice strains in the direction of applied load along RD) were determined during *in situ* uniaxial tensile test at the ambient temperature. The measurements were made at a series of applied strains after stabilisation of the load subjected to the sample. The sample strains monitored by an extensometer were held constant during the measurement intervals of 5 min. The lattice strains in the loading direction were determined for different hkl reflections and the calibration of the data for the large deformation range according to the method proposed by Baczmański *et al.* [10] was applied.

To identify the values of CRSS (critical resolved shear stresses, τ_c^{ph}) for both phases, the model predicted strains $<\varepsilon_{RD}>_{ph}$ (average values for each phase) were adjusted to the experimental ones, resulting from neutron measurements. To this end, the positions of two thresholds Γ and Ω (Fig. 2), identified respectively as yield points for the austenite and for the ferrite phases, were compared. It allowed determination of the parameters of the Voce law for each phase, independently. The relative lattice strains were shown in Fig. 2, but the initial stresses between phases were taken into account in calculations (for details of model assumptions and numerical results see [10, 11]).

A significant decrease of lattice strains in the ferritic phase and a simultaneous increase of lattice strains in the austenite phase is observed above Λ limit for the aged UR45N sample. It indicates an important relaxation of the stress in the ferritic phase, which is balanced by increasing the stress in the austenite. The observed experimental phenomenon can be predicted using our self-consistent model in which the ductile damage process is taken into account. It was assumed that at Λ threshold the damage occurs only in the ferritic phase and the results of the model were fitted to experimental $<\varepsilon_{RD}>_{ph}$ vs. Σ_{RD} plots. A good agreement between the theoretical and experimental results was obtained for most of the measured hkl reflections if the damage process was taken into account. As it is seen in Fig. 2, a decrease of the lattice strains (and corresponding stress) for the ferritic phase and an increase in the lattice strains (and corresponding stress) for the austenitic phase at Λ threshold indicate an initiation of the model-predicted damage in the ferrite.

It was found that the rate of damage, characterized by $\dot{d}^g$, is proportional to the rate of equivalent strain $\dot{\varepsilon}_{eq}^g$ (because $n^{fer}=0$ in Eq. 5). Such a stable evolution of damage in the ferritic phase is possible due to a transfer of the load into undamaged austenite, which compensates for a softening of the damaged ferritic phase. As shown in Fig. 2, a significant effect of the damage process is noticed for the Σ_{RD} stresses above Λ threshold.

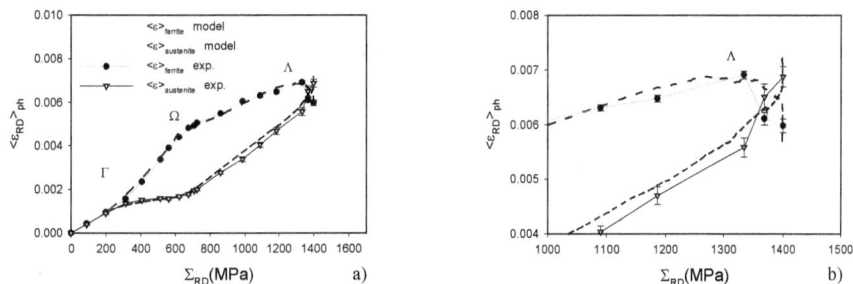

Fig. 2: Mean values of the measured elastic phase strains $<\varepsilon_{RD}>_{ph}$ vs. applied stress Σ_{RD} in the UR45N sample compared with the phase strains calculated by the self-consistent model with damage prediction (a). On the right, magnification of the range close to sample fracture (b). The thresholds Γ and Ω indicate the beginning of plasticity in austenite and ferrite, while Λ defines the initiation of the damage process.

Summary
Comparison of the elastoplastic self-consistent model with measured lattice strains allows determining the micro-mechanical properties of aluminium alloy 2124 and the Al/SiC_p composite. The partitioning of the load between metal matrix and reinforcement were correctly predicted by the model.

It was shown that the developed version of the self-consistent model could be used to predict mechanical behaviour of both phases in duplex steel as well as the consequences of damage processes occurring in the ferritic phase. The model predictions are well correlated with the results of diffraction measurements performed *in situ* during tensile test.

Acknowledgements
The work was supported partly by the NCN - Polish National Center for Science, grants: No. UMO-2011/03/N/ST8/04058, DEC-2013/11/B/ST3/03787 and partly by the MNiSW - Polish

Ministry of Science and Higher Education. Measurements at the ISIS (UK) neutron source were funded by a beamtime allocation (RB820145) from the STFC. Experiment in FLNP (JINR, Russia) was supported by beamtime allocations: 2014-10-14-18-51-38, 2015-04-18-15-20-06, 2016-04-14-19-53-04.

References

[1] J.D. Eshelby, The determination of the elastic field of an ellipsoidal inclusion and related problems, Proc. R. Soc. Lond. A 241 (1957) 376-396. https://doi.org/10.1098/rspa.1957.0133

[2] R. Hill, Continuum micromechanics of elastoplastic polycrystals, J. Mech. Phys. Solids 13 (1965) 89–101. https://doi.org/10.1016/0022-5096(65)90023-2

[3] P.A. Turner and C.N. Tomé, A study of residual stresses in Zircaloy-2 with rod texture, Acta Metall. Mater. 42 (1994) 4143–4153. https://doi.org/10.1016/0956-7151(94)90191-0

[4] B. Clausen, T. Lorentzen, M.A.M. Bourke and M.R. Daymond, Lattice strain evolution during uniaxial tensile loading of stainless steel, Mater. Sci. Eng. A 259 (1999) 17–24. https://doi.org/10.1016/S0921-5093(98)00878-8

[5] M.R. Daymond and H.G. Priesmeyer, Elastoplastic deformation of ferritic steel and cementite studied by neutron diffraction and self-consistent modelling, Acta Mater. 50 (2002) 1613–1623. https://doi.org/10.1016/S1359-6454(02)00026-5

[6] C.J. Neil, J.A. Wollmershauser, B. Clausen, C.N. Tomé and S.R. Agnew, Modeling Lattice Strain Evolution at Finite Strains. Model Verification for Copper and Stainless Steel Using in-situ Diffraction Measurement, Int. J. Plasticity 26 (2010) 1772-1791. https://doi.org/10.1016/j.ijplas.2010.03.005

[7] P. Lipinski and M. Berveiller, Elastoplasticity of micro-inhomogeneous metals at large strains, Int. J. Plastic. 5 (1989) 149–172. https://doi.org/10.1016/0749-6419(89)90027-2

[8] A. Baczmański, R. Levy-Tubiana, M.E. Fitzpatrick and A. Lodini, Elastoplastic deformation of Al/SiCp metal–matrix composite studied by self-consistent modelling and neutron diffraction, Acta Mater. 52 (2004) 1565–1577. https://doi.org/10.1016/j.actamat.2003.12.002

[9] G. Franz, F. Abed-Meraim and M. Berveiller, Effect of microstructural and morphological parameters on the formability of BCC metal sheets, Steel Res. Int. 85 (2014) 980-987. https://doi.org/10.1002/srin.201300166

[10] A. Baczmański, L. Le Joncour, B. Panicaud, M. Francois, C. Braham, A.M. Paradowska, S. Wroński, S. Amara and R. Chiron, Neutron time-of-flight diffraction used to study aged duplex stainless steel at small and large deformation until sample fracture, J. Appl. Cryst. 44 (2011) 966-982. https://doi.org/10.1107/S0021889811025957

[11] A. Baczmański, Y. Zhao, E. Gadalińska, L. Le Joncour, S. Wroński, C. Braham, B. Panicaud, M. François, T. Buslaps and K. Soloducha, Elastoplastic deformation and damage process in duplex stainless steels studied using synchrotron and neutron diffractions in comparison with a self-consistent model, Int. J. Plast. 81 (2016) 102–122. https://doi.org/10.1016/j.ijplas.2016.01.018

[12] K. Saanouni and A. Abdul-Latif, Micromechanical Modeling Of Low-Cycle Fatigue Under Complex Loadings -Theoretical Formulation, Int. J. Plasticity 12 (1996) 1111-1121. https://doi.org/10.1016/S0749-6419(96)00043-5

[13] L. L. Snead, T. Nozawa, Y. Katoh, T.-S. Byun, S. Kondo and D.A Petti, Handbook of SiC properties for fuel performance modeling, J. Nucl. Mater. 371(2007) 329–377. https://doi.org/10.1016/j.jnucmat.2007.05.016

Deformation & Modelling

MECA SENS 2017   Materials Research Forum LLC
Materials Research Proceedings **4** (2018) 11-16   doi: http://dx.doi.org/10.21741/9781945291678-2

Examination of Deformation Mechanisms of Magnesium AZ31: *in situ* X-Ray Diffraction and Self-Consistent Modelling

M. Wronski[1,a], K. Wierzbanowski[1,b*], A. Baczmanski[1,c], S. Wronski[1,d],
M. Wojtaszek[1,e], R. Wawszczak[1,f] and M. Muzyka[1,g]

[1]AGH University of Science and Technology, Faculty of Physics and Applied Computer Science,
al. Mickiewicza 30, 30-059 Kraków, Poland

[a]mwronskii@gmail.com, [b]wierzbanowski@fis.agh.edu.pl, [c]baczmanski@fis.agh.edu.pl,

[d]wronski.sebastian@gmail.com, [e]mich.wojtaszek@gmail.com,

[f]wawszczak@fis.agh.edu.pl, [g]maciejmuzyka@gmail.com

Keywords: Magnesium AZ31, Slip Systems, X-ray Diffraction, Texture, Self-consistent Model

Abstract. Samples of rolled commercial AZ31 magnesium were stretched along RD and *in situ* X-ray measurements were performed. The macroscopic stress-strain curves were examined in order to find basic mechanical material parameters. The lattice strains for different *hkl* reflections were determined as a function of the applied tensile stress. The obtained experimental results were compared with the predictions of the elasto-plastic self-consistent model. It was found that mainly slip systems are responsible for the observed material deformation and their critical resolved shear stresses (CRSS) were evaluated.

Introduction

One observes a growing interest in magnesium and its alloys in the last decade (e.g., [1]). Magnesium due to its unique properties, finds many technological application. It has an advantageous ratio of the yield strength and mass density. The disadvantage of this material is a low ductility at room temperature, which severely limits its formability. Therefore, it is necessary to better understand its deformation mechanisms, which can help to optimize its formability. The aim of this work was an examination of the mechanisms of plastic deformation of magnesium AZ31 basing on X-ray studies carried out during in situ tensile tests. The tensile force was applied along the rolling direction of the initial material. In this experiment two kinds of data were obtained: crystal lattice strains for different reflections and the macroscopic stress-strain curve. In order to determine selected material parameters the experimental results were compared with deformation model predictions. The elasto-plastic self-consistent model was used in the present study [2, 3]. The slip and twinning systems, which were taken into account in the calculations are listed in Tables 1 and 2.

Generally, a comparison of experimental and calculated lattice strains vs. applied load allows the identification of active deformation mechanisms of the examined material, i.e., slip and twinning systems and the estimation of their critical resolved shear stresses (CRSS). Such a study can also provide information on the activity of deformation mechanisms and their influence on the observed material characteristics.

Table 1. Slip systems of magnesium [4,5] and values of their critical resolved shear stresses (CRSS), which led to the best agreement between measured and predicted lattice strains.

Slip system	Miller indices	CRSS [MPa]
Basal B	$\{0001\}\langle11\bar{2}0\rangle$	17
Prismatic P	$\{1\bar{1}00\}\langle11\bar{2}0\rangle$	75
Pyramidal π<a>	$\{1\bar{1}01\}\langle11\bar{2}0\rangle$	75
Pyramidal π_1 <c+a>	$\{0\bar{1}11\}\langle11\bar{2}3\rangle$	80
Pyramidal π_2 <c+a>	$\{11\bar{2}2\}\langle11\bar{2}3\rangle$	70

Table 2. Twinning systems of magnesium.

Twinning systems	Miller indices
Compression twins	$\{11\bar{2}2\}\langle\bar{1}\bar{1}23\rangle$
Tensile twins	$\{10\bar{1}2\}\langle\bar{1}011\rangle$

Experimental

The initial samples were cut along the rolling direction (RD) from the commercial AZ31 sheet, with chemical composition shown in Table 3. Its microstructure was examined by the EBSD technique, using the Stereoscan Cambridge S360 microscope. The orientation map of inverse pole figure for the normal direction (ND) is shown in Fig. 1. The average grain size, as determined from EBSD measurements, was around 12 μm. Texture of the initial material contained a predominating $< 0001 > \parallel RD$ fibre component, which is clearly visible on the $\{0001\}$ pole figure – Fig. 2. This strong texture implies a predominating orientation of hexagonal crystal unit cells shown schematically in Fig. 3.

Table 3. Chemical composition of AZ31.

Element	Al	Zn	Mn	Cu	Mg
Mass fraction [%]	$2.5 - 3.5$	$0.7 - 1.3$	$0.2 - 1.0$	0.05	$94.15 - 96.55$

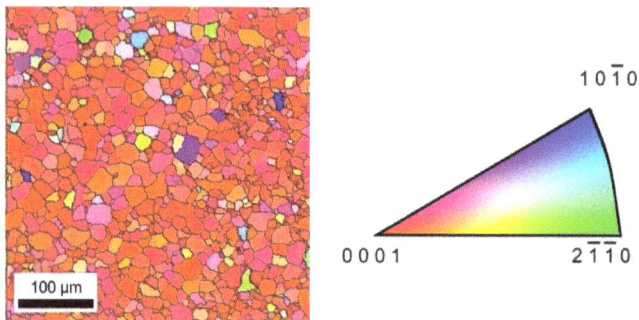

Fig.1: Initial microstructure of the examined magnesium alloy AZ31. EBSD map of inverse pole figure for the sample normal direction (ND) is shown.

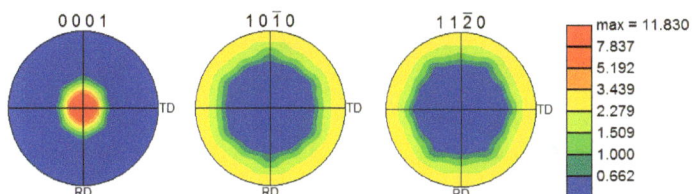

Fig. 2: *Determined pole figures for the examined magnesium sample.*

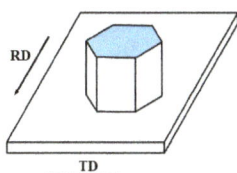

Fig. 3: *Predominating position of lattice cells of grains expressed in the sample coordinate system defined by the primary rolling process (RD and TD – rolling and transverse directions). The majority of crystallites exhibit orientations rotated around c-axis corresponding to an <0001> axial texture.*

The scheme of the diffraction experiment is shown in Fig. 4. The X-ray in-situ measurements were done during tensile tests.

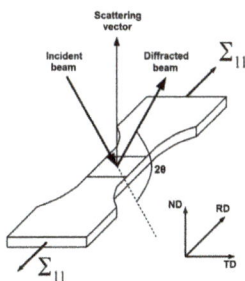

Fig. 4: *X-ray diffraction geometry used for in-situ measurements during tensile tests.*

Results and discussion

The measured and predicted macroscopic stress-strain curves are shown in Fig. 5. A perfect agreement is obtained between the experiment and model calculation. The elasto-plastic transition occurs in the range between 130 MPa and 140 MPa.

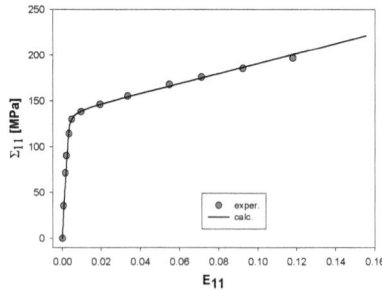

Fig. 5: Comparison between experimental and calculated (self-consistent elasto-plastic model) stress-strain curves.

A more detailed insight into material structure can be done by internal stress measurements (e.g., [6-9]). The diffraction in-situ experiments were performed using X-ray Cu radiation and the analysed *hkl* reflections are listed in Table 4.

Table 4. {hkl} reflection used to examine the AZ31 sample.

h	0	1	1	1	0	1	1	1	0	2	1
k	0	0	0	0	0	0	1	0	0	0	0
l	2	1	2	3	4	4	4	5	6	5	6

The lattice strains were measured along the normal sample direction (ND), i.e., perpendicular to the applied force. The determined lattice strains $<\varepsilon_{33}>_{\{hkl\}}$ versus applied stress component Σ_{11}, for different *hkl* reflections, are shown in Fig. 6. The best fits of the model results to experimental curves are superimposed in these figures. The presented theoretical results were obtained with optimal values of the critical resolved shear stresses (CRSS) listed in Table 1. It is interesting to note that similar values of CRSS were found in the papers [10, 11], where neutron diffraction results (i.e., for bulk material) were modelled by the visco-plastic self-consistent model with Voce law for slip systems hardening. Also the results of the present authors, obtained using neutron diffraction and the elasto-plastic deformation model are similar.

It can be noted that for reflections {102}, {103}, {114}, {205} some non-linearity is observed in the range of Σ_{11} values up to 140 MPa (in principle this is elastic range – according to macroscopic stress-strain curve). This behaviour can be explained by an early activations of selected slip systems (e.g., of basal systems), caused by the presence of residual stresses originating from the preceding rolling process. In the future work this effect should be taken into account and checked by a prior modelling.

In other groups of grains, such as by {101}, {002}, {106}, {104}, {004}, {105},{006} reflections, regular linear relations appear in the elastic range. At Σ_{11}=140 MPa one observes inflection points, which confirms the transition to the plastic range (activation of consecutive slip systems). It can be generally concluded that in all the cases the trends of experimental curves are approximately reproduced by the model predictions.

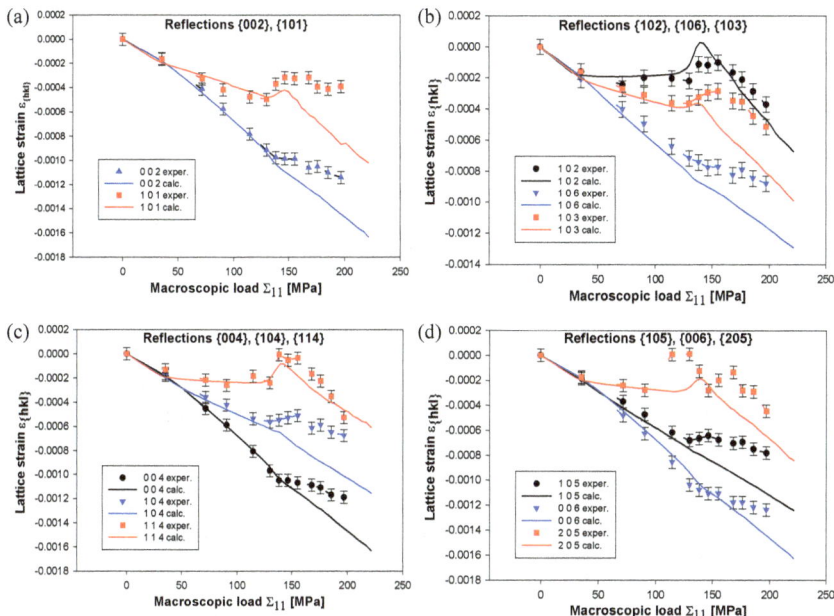

Fig. 6: Lattice strains ($<\varepsilon_{33}>\{hkl\}$) in the direction perpendicular to the applied force (i.e., along ND) versus applied stress component $\sum_{11}$: experimental results (broken lines with points) are compared with model predictions (solid lines). The results for the following reflections are shown: (a) {002}, {101}; (b) {102}, {106}, {103}; (c) {004}, {104}, {114}; (d) {105}, {006}, {205}.

The influence of twinning on the predicted lattice strains was also examined in the frame of a deformation model. The estimated CRSS for the tensile and compressive twinning systems are 50 MPa and 60 MPa, respectively(CRSS for tensile twins found in [10,11] is at similar level).

Introduction of this mechanism improves only slightly the agreement between experimental and model results in the case of basic reflection {002} (and also the higher order reflections {004} and {006} corresponding to the same crystallographic plane). In the case of other reflections, e.g., {102}, it led to worse agreement. Consequently, taking into account only slip systems was sufficient to find the best fit of experimental lattice strains to the calculated ones.

It should be mentioned that the experimentally determined twin volume in magnesium deformed by tension varies between 0 and 4 % for average grain size in the range of 10-20 μm [12]. It is a reason why inclusion of twinning in the calculations did not improve the agreement between experimental and predicted lattice strains.

Conclusions

The present study of deformation mechanisms of magnesium AZ31 was based on the comparison of measured lattice strains with those predicted by the elasto-plastic self-consistent deformation model. The optimal CRSS values, leading to the best agreement between experimental and theoretical results, were evaluated. Moreover, it was found that slip systems play a predominant role in the plastic deformation of magnesium. Inclusion of tensile and compression twinning systems in the calculations did not generally improve the agreement between predicted and experimental results.

Acknowledgments

This study was financed by grants of the Polish National Centre for Science (NCN) No: DEC-2015/19/D/ST8/00818 and DEC-2013/11/B/ST3/03787 and by the AGH statutory works No. 11.11.220.01.

References

[1] M. Easton, A. Beer, M. Barnett, C. Davies, G. Dunlop, Y. Durandet, S. Blacket, T. Hilditch and P. Beggs, Magnesium alloy applications in automotive structures, JOM 60 (2008) 57–62. https://doi.org/10.1007/s11837-008-0150-8

[2] K. Wierzbanowski, A. Baczmanski, P. Lipinski and A. Lodini, Elasto-plastic models of polycrystalline material deformation and their applications, Arch. Metall. Mater. 52 (2007) 77-86

[3] A. Baczmański, N. Hfaiedh, M. François and K. Wierzbanowski, Plastic Incompatibility Stresses and Stored Elastic Energy in Plastically Deformed Copper, Mat. Sci. Eng. A. 501 (2009) 153-165. https://doi.org/10.1016/j.msea.2008.09.072

[4] Y.N. Wang and J.C. Huang, Texture analysis in hexagonal materials, Mater. Chem. Phys. 81 (2003) 11-26. https://doi.org/10.1016/S0254-0584(03)00168-8

[5] M.J. Philippe, M. Serghat, P. Van Houtte and C. Esling, Modelling of texture evolution for materials of hexagonal symmetry—II. application to zirconium and titanium α or near α alloy, Acta Metall. Mater. 43 (1995) 1619- 1630. https://doi.org/10.1016/0956-7151(94)00329-G

[6] S. Wronski, M. Wrobel, A. Baczmanski and K. Wierzbanowski, Effects of cross-rolling on residual stress, texture and plastic anisotropy in f.c.c. and b.c.c. metals, Mater. Charact. 77 (2013) 116-126. https://doi.org/10.1016/j.matchar.2013.01.005

[7] K. Wierzbanowski, J. Tarasiuk, B. Bacroix, K. Sztwiertnia and P. Gerber, Recrystallization Textures – Two Types of Modelling, Met. Mater. Int. 9 (2003) 9-14. https://doi.org/10.1007/BF03027223

[8] S. Wronski, K. Wierzbanowski, B. Bacroix, M. Wróbel, E. Rauch, F. Montheillet and M. Wroński, Texture heterogeneity of asymmetrically rolled low carbon steel, Arch. Metall. Mater. 54 (2009) 89-102.

[9] M. Marciszko, A. Baczmanski, K. Wierzbanowski, M. Wróbel, C. Braham, J.-P. Chopart, A. Lodini, J. Bonarski, L. Tarkowski and N. Zazi, Application of multireflection grazing incidence method for stress measurements in polished Al–Mg alloy and CrN coating, Appl. Surf. Sci. 266 (2013) 256-267. https://doi.org/10.1016/j.apsusc.2012.12.005

[10] B. Clausen, C.N. Tome, D.W. Brown and S.R. Agnew, Reorientation and stress relaxation due to twinning: Modeling and experimental characterization for Mg, Acta Mater. 56 (2008) 2456–2468. https://doi.org/10.1016/j.actamat.2008.01.057

[11] S.R. Agnew, C.N. Tome, D.W. Brown, T.M. Holden and S.C. Vogel, Study of slip mechanisms in a magnesium alloy by neutron diffraction and modeling, Scripta Mater. 48 (2003) 1003–1008. https://doi.org/10.1016/S1359-6462(02)00591-2

[12] A. Jain, O. Duygulu, D.W. Brown, C.N. Tome and S.R. Agnew, Grain size effects on the tensile properties and deformation mechanisms of a magnesium alloy, AZ31B, sheet, Mat. Sci. Eng. A. 486 (2008) 545–555. https://doi.org/10.1016/j.msea.2007.09.069

MECA SENS 2017
Materials Research Proceedings 4 (2018) 17-22

Materials Research Forum LLC
doi: http://dx.doi.org/10.21741/9781945291678-3

In situ Synchrotron X-ray Measurement of Strain Fields near Fatigue Cracks grown in Hydrogen

M. Connolly[1,a*], P. Bradley[1,b], A. Slifka[1,c], D. Lauria[1,d] and E. Drexler[1,e]

[1]National Institute of Standards and Technology, 325 Broadway Boulder, CO 80305 USA

[a]matthew.connolly@nist.gov, [b]peter.bradley@nist.gov, [c]andrew.slifka@nist.gov, [d]damian.lauria@nist.gov, [e]elizabeth.drexler@nist.gov

Keywords: Strain, Hydrogen, Fracture, Fatigue, Synchrotron, Diffraction

Abstract. The embrittlement and enhanced fatigue crack growth rate of metals in the presence of hydrogen is a long-standing problem [1-5]. In an effort to determine the dominant damage mechanism behind hydrogen-assisted fatigue crack growth, we performed high-energy X-ray diffraction (HEXRD) measurements to characterize the strain fields near cracks grown both in air, as well as in a hydrogen environment. An enhancement in the magnitude and spatial extent of the strain field near the crack grown in hydrogen compared with the strain field near the crack grown in air was observed. We discuss the differences between the measured in-air and in-hydrogen crack-tip strain fields in the context of the two leading damage mechanisms proposed in the literature.

Introduction

Proposed mechanisms of hydrogen embrittlement include hydrogen-enhanced decohesion (HEDE) and the hydrogen-enhanced localized plasticity (HELP) [6-10]. In the HEDE mechanism, decohesion occurs either through a weakening of Fe-Fe bonds ("intra-lattice" decohesion) or from a buildup of hydrogen at grain boundaries ("inter-lattice decohesion"). In the HELP mechanism, the introduction of hydrogen gas leads to failure through localized plastic deformation from enhancement of dislocation mobility in the steel framework. Quantifying strain fields from fatigue cracking is crucial to the understanding of the underlying mechanism(s) behind hydrogen embrittlement (HE) and to the study of HE and hydrogen-assisted fatigue crack growth rate (HA-FCGR). For example, a measurement of the strain field ahead of a crack tip can directly confirm predictions from the HEDE mechanism of intra-lattice decohesion.

The synchrotron source available at Argonne National Laboratory's Advanced Photon Source (APS) and its HEXRD technique on the 1-ID beamline are useful tools for probing the strain field ahead of cracks grown in hydrogen. The use of a 2D detector and short X-ray wavelengths allows for the simultaneous measurement of the two in-plane strain components in the planar, compact tension (C(T)) specimen geometry commony used for FCGR measurements. The load frame available at APS for mechanical testing is capable of fatigue cycling up to 10 Hz, which allows a full FCGR test to be performed within a single awarded beam time. Further, high energy X-rays have high flux and high penetration through sample chamber materials, which allow for *in situ* measurements. For HE and HA-FCGR studies, *in situ* measurements are crucial because of the rapid diffusion of hydrogen out of ferritic steels. Even after extended exposure to hydrogen, the deleterious effect of hydrogen on ferritic steels after the specimen has been removed from the hydrogen environment for just a short period of time (~ 15 min) [7]. Therefore, in order to fully understand the HA-FCGR mechanism, it is necessary to perform any measurements *in situ*.

MECA SENS 2017
Materials Research Proceedings **4** (2018) 17-22

Materials Research Forum LLC
doi: http://dx.doi.org/10.21741/9781945291678-3

Materials

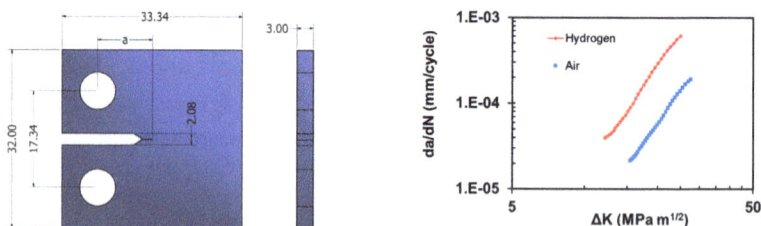

Fig. 1: Geometry of the X100 steel CT specimen used for HEXRD measurements (left). Units are in mm. FCGR curves for the 4130 steel showing a significant enhancement in FCGR in hydrogen compared to in air (right).

The material used for this study was a 4130 steel in a compact tension (CT) geometry, with dimensions as shown in Fig. 1 (left). These type of quench and temper steels exhibit a drastic increase in FCGR for fatigue cracks grown in hydrogen compared to those grown in air (Fig. 1, right).

Experimental Methods

The X-ray measurements were performed with a chamber designed for use in neutron and X-ray scattering measurements of metallic specimens under mechanical load in up to 1.7 MPa gaseous atmosphere [11]. At beamline 1-ID [12, 13], X-rays produced by the synchrotron source are monochromated with a bent Si(111) crystal. The beam is then narrowed by the beam slit, which results in a minimum beam size of 20 μm × 20 μm. For experiments that use the hydrogen chamber, the chamber is mounted to the mechanical load frame, and the load frame is mounted on a stage capable of translations perpendicular to the beam in the horizontal and vertical directions. Five helium purges and five hydrogen purges were performed at pressures of 1.7 MPa to clean the chamber prior to fatigue testing. Commerical purity helium (99.995 % pure) and research-grade hydrogen (99.9995 %) were used. The pressure was maintined at 1.7 MPa throughout the course of the X-ray measurements.

The X-ray measurements were performed with 80.725 keV (λ = 0.15359 Å) X-rays. The X-ray beam size was 20 μm × 20 μm, which determines the spatial resolution of the strain mapping. For the diffraction measurements, a pixelated area detector with a pixel pitch of 200 μm and 2048 × 2048 pixel array was used. Radiographs were collected with a camera that has an effective pixel size of 1 μm after lensing. The radiograph detector was positioned between the sample and diffraction detector during radiograph measurements, and was moved out of the beam by a translation stage for diffraction measurements. The mechanical load frame was a modified servo-hydraulic system with a force capacity of +/- 15 kN and was equipped with a 2.5 kN load cell, mounted above the chamber. A 44 kN load cell was mounted inside the sample chamber at the base of the load train. A crack mouth opening displacement (CMOD) gauge was attached to the CT specimen at the load line. Fatigue cycling was performed in displacement control with a frequency of 1 Hz. To determine the unstrained lattice spacing, the lattice spacing was measured on an illuminated spot far from the crack tip when the specimen was under zero load. By using as a reference parameter the lattice spacing far from the crack tip while the material was in air, the reported strain measurements then incorporate both the effect of interstitial hydrogen [14] as well as the response of the applied stress, to describe the full deformation at the crack tip. In the unloaded specimen, the variance in the lattice spacing

MECA SENS 2017   Materials Research Forum LLC
Materials Research Proceedings **4** (2018) 17-22   doi: http://dx.doi.org/10.21741/9781945291678-3

throughout the scanned material was on the order of 10^{-4}, leading to a systematic uncertainty in the strain of 50 μstrain.

Results and Discussion

Crack tip radiographs. Radiographs of the crack grown in hydrogen, after N = 20,000 cycles, N = 38,000 cycles, and N = 54,000 cycles, and in air after N = 0, N = 100,000, and N = 149,000 cycles with a load of 1700 N applied are shown in Fig. 2.

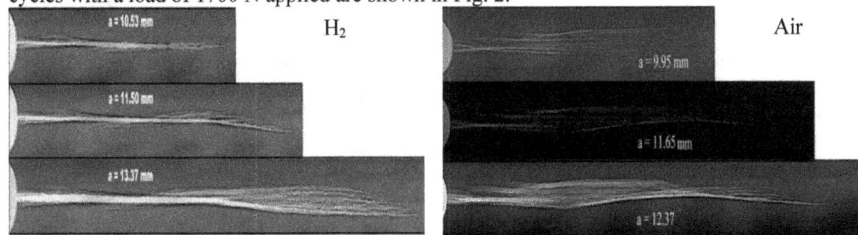

Fig. 2: X-ray radiographs of the crack tip with an applied force of 1700 N after N = 20,000 cycles, N = 38,000 cycles, and N = 54,000 cycles grown in hydrogen, and for N = 0 cycles, N = 100,000 cycles, and N = 149,000 cycles in air (from top to bottom).

For the crack grown in H_2 the crack lengths shown in Fig. 2 (left) correspond to ΔK = 12.28 MPa $\sqrt{m}$, ΔK = 13.54 MPa $\sqrt{m}$, and ΔK = 16.58 MPa $\sqrt{m}$, which reside at the start, middle, and end of the "knee" in the hydrogen da/dN curve shown in Fig. 1. Likewise, the air cracks shown in Fig. 2 (right) correspond to ΔK = 11.7 MPa $\sqrt{m}$, ΔK = 13.7 MPa $\sqrt{m}$, and ΔK = 14.9 MPa $\sqrt{m}$. The 2D radiographs are a projection of the 3D attenuation coefficient summed through the thickness of the specimen. The multiple crack branches observed in each image therefore correspond to the crack growing in different planes. Similar crack branching has been observed using synchrotron radiography in a fretting crack grown in a 2024 aluminum alloy [15]. In the first two images of the crack grown in hydrogen, some crack branching is observed behind the tip of the crack, however the crack tip is much sharper than the final hydrogen crack image or any of the air crack images. It is interesting to note the crack grown in air exhibits far greater crack branching when the crack is short, whereas the hydrogen crack does not exhibit secondary damage to the same extent until the crack is longer (corresponding to higher ΔK).

Fig. 3 shows the loading direction strain (ε_{yy}) as a function of distance ahead of the crack tip, r, for the six crack conditions shown in Fig. 2. Closest to the crack, the expected $r^{-1/2}$ strain dependence is observed until ~ 3 mm ahead of the crack tip. Far from the crack tip in the longest cracks, compressive strain is observed attributable to hinging of the CT specimen. In the shorter cracks (< 11 mm) the strain field forms a sharp cusp at $r = 0$. In the longer cracks with $P = 850$ N, the strain magnitude increases moving away from the crack tip, until it reaches a maximum at ~ 0.25 mm, then decreases with $r^{-1/2}$ dependence. The lowered elastic strain nearest the crack tip is likely from the build up of plastic deformation during fatigue cycling. The magnitude of the drop in strain at the crack tip is significantly higher in the crack grown in hydrogen compared with that of the crack grown in air (400 μstrain compared with 100 μstrain). This may suggest additional plastic deformation localized at the crack tip from the presence of hydrogen. Enhanced near-crack plastic deformation is consistent with the HELP mechanism. The extent of the plastic deformation can be quantified in the dislocation density that is determined through peak broadening analysis; however, the current work focuses on the elastic strain only.

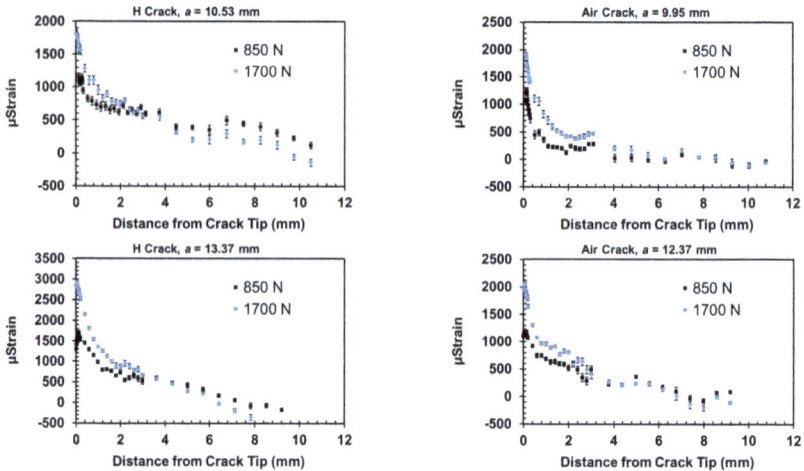

Fig. 3: Loading direction (ε_{yy}) strain values near cracks grown in hydrogen (left) and in air (right), with 1700 N and 850 N of applied load.

In comparing the hydrogen and air cracks, both the maximum strain magnitude is enhanced for the crack grown in hydrogen compared with that grown in air. The rate at which the strain drops as a function of distance from the crack is lower for the crack grown in hydrogen. These differences are shown most predominately in Fig. 4, which shows the crack-tip strain values in the loading direction ahead of cracks grown in air compared with one grown in hydrogen with the same crack length of $a = 11.5$ mm. The largest difference between the two strain fields are for $r < 2$ mm ahead of the crack tip. The crack strain field in air decays very sharply to within 25 % of its maximum by $r = 2$ mm ahead of the crack tip, whereas the decay of the crack strain field in hydogen is much shallower, reaching 25 % of its maximum near $r = 3$ mm. Beyond this region, the strain fields are nearly independent of applied load.

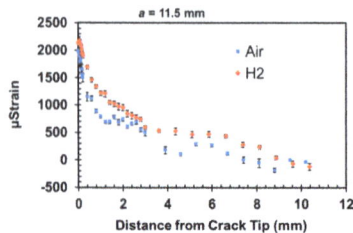

Fig. 4: Strain fields (ε_{yy}) corresponding to the loading direction near cracks grown in hydrogen and air for an 11.5 mm crack.

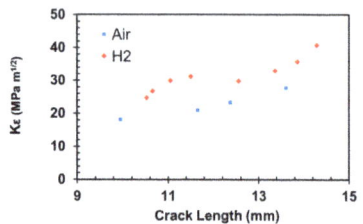

Fig. 5: Strain-based crack tip deformation parameter K_ε as a function of crack length for cracks grown in air and in hydrogen.

Althrough the strain fields were measured under static loads, we can relate the differences in static crack-tip strain fields to measured FCGR. For r within the "K-dominant region"[16] (that is, for r greater than the size of the plastic zone but sufficiently far from the specimen edge to avoid significant influence from the specimen boundary), the K parameter completely defines the strain field ahead of the crack tip. Visual inspection of the measured strain fields indicates the K-dominant region to be approximately 0.5 mm $< r <$ 3 mm. Based on the strain fields measured here, the use of the ΔK prescribed by ASTM E647 [17], the standard test method for measuring FCGRs, does not accurately describe either the in-air and in-hydrogen fatigue cracks. Therefore the independent variable in the FCGR curves must be modified in order to correctly normalize the data with respect to the true deformation state at the crack tip. These data suggests the correct normalization should be done with respect to the strain rather than the applied stress state as in the ΔK formulation in ASTM E-647. Here we define a new parameter K_ε which satisfies

$$\sigma_{ij}(r,\theta) = \frac{K_\varepsilon}{\sqrt{2\pi r}} f_{ij}(\theta), \tag{1}$$

where r and θ are the distance from the crack tip and angle with respect to the crack, respectively, and $f_{ij}(\theta)$ is a geometrical factor. Assuming plane-stress conditions the loading-direction strain, ϵ_{yy}, is

$$\epsilon_{yy} = \frac{1}{E}(-\nu\,\sigma_{xx} + \sigma_{yy}), \tag{2}$$

where ν is Poisson's ratio and E is Young's modulus. With $\theta = 0$, substituting Eq. (1) into (2) gives:

$$\epsilon_{yy} = \frac{(1-\nu)}{E}\left(\frac{K_\varepsilon}{\sqrt{2\pi r}}\right). \tag{3}$$

With Eq. (3), we fit the measured strain fields ϵ_{yy} at $\theta = 0$ to determine the strain-based K_ε parameter for each environmental condition and crack length. Fig. 5 shows K_ε as a function of crack length for the cracks grown in air and in hydrogen with an applied load of $P = 1700$ N. For each crack extension, K_ε for the crack grown in hydrogen is larger than that for the crack grown in air by ~ 50 %.

Conclusion
We have shown through *in situ* HEXRD strain mapping an enhancement in the magnitude of the crack-tip strain attributable to the presence of hydrogen. The results presented here are consistent with the intra-lattice HEDE mechanism. The enhancement suggests that a strain-based parameter is better suited for hydrogen FCGR measurements in order to accurately describe the crack-tip deformation state and for comparing air and hydrogen FCGRs. For the 4130 steel and crack lengths measured here, we observe a ~50 % enhancement of the strain-based parameter in hydrogen compared with that in air. Radiographs of the crack tip show significant differences between air and hydrogen cracks.

References
[1] J.R. Fekete, J.W. Sowards and R.L. Amaro, Economic impact of applying high strength steels in hydrogen gas pipelines, Int. J. Hydrog. Energy. 40(33) (2015) 10547-10558. https://doi.org/10.1016/j.ijhydene.2015.06.090

[2] W.C. Leighty, J. Holloway, R. Merer, G. Keith and D.E. White, Compressorless hydrogen transmission pipelines deliver large-scale stranded renewable energy at competitive cost, Proceedings of the 23rd World Gas Conference (2006)

[3] A.J. Slifka, E.S. Drexler, N.E.Nanninga, Y.S. Levy, J.D. McColskey, R.L. Amaro and A.E. Stevenson, Fatigue crack growth of two pipeline steels in a pressurized hydrogen environment, Corros. Sci. 78 (2014) 313-321. https://doi.org/10.1016/j.corsci.2013.10.014

[4] N.E. Nanninga, Y.S. Levy, E.S. Drexler, R.T. Condon, A.E. Stevenson and A.J. Slifka, Comparison of hydrogen embrittlement in three pipeline steels in high pressure gaseous hydrogen environments, Corros. Sci. 59 (2012) 1-9. https://doi.org/10.1016/j.corsci.2012.01.028

[5] R.L. Amaro, N. Rustagi, K.O. Findley, E.S. Drexler and A.J. Slifka, Modeling the fatigue crack growth of X100 pipeline steel in gaseous hydrogen, Int. J. Fatigue. 59 (2014) 262-271. https://doi.org/10.1016/j.ijfatigue.2013.08.010

[6] R.A. Oriani, A mechanistic theory of hydrogen embrittlement of steels, Berichte der Bunsengesellschaft für physikalische Chemie 76(8) (1972) 848-857.

[7] P.P. Darcis, J.D. McColskey, A.N. Lasseigne and T.A. Siewert, Hydrogen effects on fatigue crack growth rate in high strength pipeline steel, Effects of Hydrogen on Materials: Proceedings of the 2008 International Hydrogen Conference (2009) 381.

[8] T. Tabata and H.K Birnbaum, Direct observations of the effect of hydrogen on the behavior of dislocations in iron, Scr. Metall. 17(7) (1983) 947-950. https://doi.org/10.1016/0036-9748(83)90268-5

[9] H.K. Birnbaum and P. Sofronis Hydrogen-enhanced localized plasticity—a mechanism for hydrogen-related fracture, Mater. Sci. Eng. A 176(1-2) (1994) 191-202. https://doi.org/10.1016/0921-5093(94)90975-X

[10] I.M. Robertson, P. Sofronis, A. Nagao, M.L. Martin, S. Wang, D.W. Gross and K.E. Nygren, Hydrogen embrittlement understood, Metall. Mater. Trans. A 46(6) (2015) 2323-2341. https://doi.org/10.1007/s11661-015-2836-1

[11] M.J. Connolly, P.E. Bradley, A.J. Slifka and E.S. Drexler, Chamber for mechanical testing in H2 with observation by neutron scattering, Rev. Sci. Instrum. 88(6) (2017) 063901. https://doi.org/10.1063/1.4986471

[12] J. Almer, Advanced Photon Source. Retrieved July 20, 2017, from https://www1.aps.anl.gov/sector-1/1-id

[13] A.S. Gill, Z. Zhou, U. Lienert, J. Almer, D.F Lahrman, S.R. Mannava and V.K. Vasudevan, High spatial resolution, high energy synchrotron X-ray diffraction characterization of residual strains and stresses in laser shock peened Inconel 718SPF alloy, J. Appl. Phys. 111(8) (2012) 084904. https://doi.org/10.1063/1.3702890

[14] G.L. Nash, H. Choo, P. Nash, L.L. Daemen and A.M. Bourke, Lattice Dilation in a Hydrogen Charged Steel, International Centre for Diffraction Data. Adv. X-Ray Anal. (2003) 238-239.

[15] H. Proudhon, J.Y. Buffière and S. Fouvry, Three-dimensional study of a fretting crack using synchrotron X-ray micro-tomography, Eng. Fract. Mech. 74(5) (2007) 782-793. https://doi.org/10.1016/j.engfracmech.2006.06.019

[16] Anderson, T. L. (2017). Fracture mechanics: fundamentals and applications, second ed., CRC press, Boca Raton, 1995.

[17] ASTM International, Standard test method for measurement of fatigue crack growth rates. (2011)

MECA SENS 2017
Materials Research Proceedings **4** (2018) 23-28

Materials Research Forum LLC
doi: http://dx.doi.org/10.21741/9781945291678-4

Stress Evaluation of Adhesively Bonded Lap Joints with Aluminum 2024-T3 Adherents Using FEA

M. Khodja[1,2,a*], A. Ahmed[1,b], F. Hamida[1,c], G. Corderley[2,d] and S. Govender[2,e]

[1] LMPM, Mechanics and physics of materials Laboratory, Mechanical Engineering Department, University of Djillali Liabes, Sidi Bel Abbes, BP 98 Cité ben M'Hidi Sidi Bel Abbes 22000, Algeria

[2] CSIR Materials Science and Manufacturing, Meiring Naude Road, Pretoria, 0184, South Africa

[a]MKhodja@csir.co.za, [a]khodja.malika7@gmail.com, [b]amiri_ahm@yahoo.fr, [c]hamida.fekirini@univ-sba.dz, [d]GCorderl@csir.co.za, [e]SGovender@csir.co.za

Keywords: Adhesive Bonded Joints, Finite element analysis (FEA), Simple-Lap Joint (SLJ), Single Step-Lap Joints (SSLJ), Shear Stresses, Peel Stresses, Von Mises Stresses

Abstract The objective of this study is to develop a numerical approach which will lead to a method to aid in the design of bonded assemblies. An AA2024-T3 aluminum alloy was used as adherent for this study with Adekit-140 as the adhesive. The overlapping surfaces of the adherent and the adhesive were modelled with 3D models that were based on surface-to-surface contact elements. Analyses were performed where the length and the thickness of overlap were fixed, keeping the bonding area the same for all geometries. Peel stresses developing at the edges of the overlap area of the adhesively bonded single lap joints subjected to static tensile loading have a profound effect on the damage of the joint. The reduction in the stress values formed at the edges of the overlap area or the transfer of these stresses to the middle part of the overlap area increase the strength of the joint. It was noted that there is symmetry in the stress distribution about the middle of adhesive joint layer according to the length of overlap region. The maximum stresses were at the edge of the bond. Observations have been made on peel and shear stresses in the adhesive layer.

Introduction

Adhesively bonded joints are preferred due to their advantages such as formation of uniform stress distributions, ability to join different materials, high fatigue resistance and impermeability [1]. However, in the adhesively bonded joints, extreme levels of stress concentrations form at the edges of the overlap area, which significantly influences the strength of the joint. In order to use the adhesive bonding technique and to increase the load carrying capacity of the joint, the effect of these stresses forming at the free edges of the bonding area should be reduced. Lap joints, specifically the adhesive single-lap joint, have been studied thoroughly throughout the years. Analytic solutions date back as far as Volkersen [2] and his simplified solution in 1938 that is still more accurate than the current ASTM standards D1002-10 and D3983-98 used to determine the shear strength and shear modulus, respectively. Peel stresses developing at the edges of the overlap area of the adhesively bonded single lap joints subjected to static tensile loading have a profound effect on the damage of the joint. The reduction in the stress values formed at the edges of the overlap area or the transfer of these stresses to the middle part of the overlap area increase the strength of the joint. Results are discussed, followed by numerical work and validation.

Methods

Several factors need to be considered when designing adhesive joints, notably the peak stresses at the ends of the overlap area and the stresses due to bending moments. It is necessary to be able to determine the state of stress within the joint during the design process in order to achieve the

MECA SENS 2017   Materials Research Forum LLC
Materials Research Proceedings **4** (2018) 23-28   doi: http://dx.doi.org/10.21741/9781945291678-4

correct joint strength. This need necessitated the study of the stress state in a simple adhesive lap joint based on the analytical theories of Volkersen [2] and Goland & Riesener [3].

Model definition and validation with analytical solution

Three-dimensional finite element modeling (3D FEM). We consider a lap joints consisting of two aluminum 2024-T3 plates, E = 68800 MPa joined by an adhesive type ADEKIT A140 [4], E = 2690 MPa and Shear modulus is 1000 MPa of 0.25 mm thickness.

The adhesive and adherents are assumed to be linear elastic, homogeneous and isotropic. This adhesive is used in the aeronautical industry and has excellent mechanical and thermal performance up to 100°C, excellent strength to dynamic loads (vibrations and impacts) and it is adapted to stringent aging and aggressive environments. The dimensions of the various substrates are presented in Fig. 1 which shows the boundary conditions, mechanical properties of material used and the applied nominal stress being 25 MPa.

The model of the tapered joint has the same dimensions as the single lap joint. The thickness of the overlapping part is identical to that of the simple overlap, as well as the thickness of the adhesive. To compare the stress distribution, the other adhesive joints was subjected to the same loading as that of the single overlap joint. The thickness of the substrate of the stepped adhesive joint was fixed at 4 mm as shown in Fig. 1. Finite element analyses of lap joints were performed to calculate stresses in the joints using Abaqus software. The respective patterns of the full finite element meshes of lap joints are shown in Fig. 2. The entire specimen was modelled using an eight node quadrilateral element and the mesh refined in the adhesive layer.

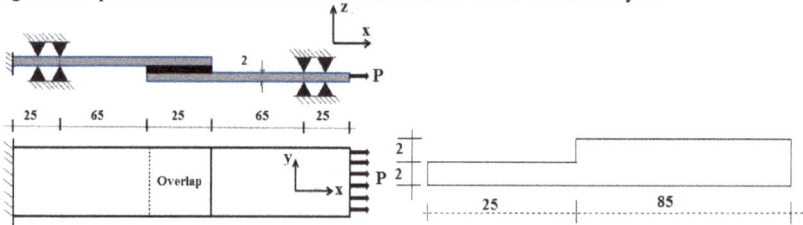

Fig. 1: Single-lap adhesively bonded joint (dimensions in mm) and Dimensions considered for single step-lap adhesively bonded joint.

In the case of analysis of adhesively bonded joints, small elements were used within the adhesive layer and around the adhesive–adherent interfaces and larger elements in the outer regions of the adherents. It is essential to model the adhesive layer with elements which are thinner than the adhesive thickness. The result is that the FE mesh must be several orders of magnitude more refined in every small region than is needed in the rest of the joint. It is also important that a smooth transition between the adherents and adhesive is provided. The finer elements mesh is refined on the two ends of the overlap length.

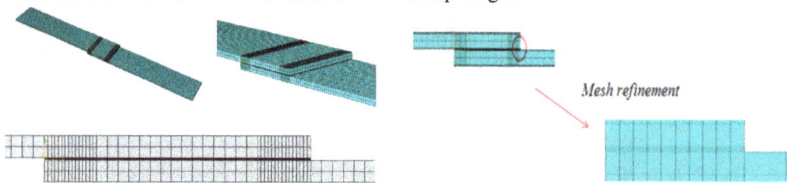

Fig. 2: Typical finite element model and mesh refinement used in lap joint analysis.

MECA SENS 2017   Materials Research Forum LLC
Materials Research Proceedings **4** (2018) 23-28   doi: http://dx.doi.org/10.21741/9781945291678-4

Analytical Evaluation of Stress Components for Simple Lap. Fig. 3 shows the comparison between the analytical method and the finite element method on the peel and shear stresses distribution. The comparison of the two models of Volkersen and Goland & Reissner for the maximum shear stress shows that the two models behave in a similar way with some variation in the predicted stresses. This variation in the stresses is probably due to the effect of the bending moment induced by the applied loads. The general results show that the analytical and FEA results were in good agreement and proved suitable for single lap joints [5].

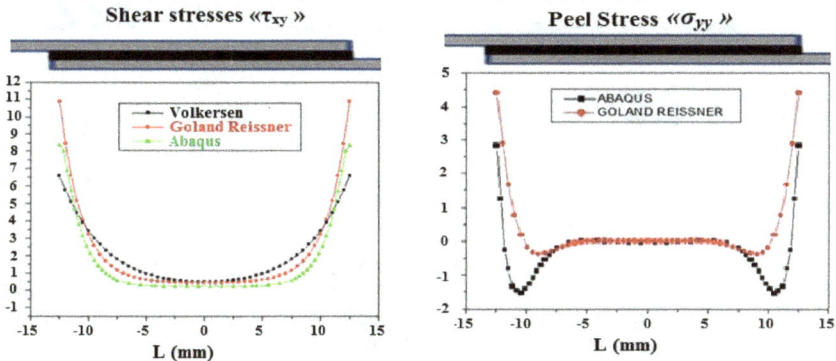

Fig. 3: Comparison between FEM and analytical method on distribution of the adhesive shear and peel stress along the overlap length [5].

Results & Discussions

Stress state in the stepped adhesive joint 1 and joint 2. After the analytical results were compared with FEA results using Abaqus [6] and validated, the general results show that the analytical and FEA results were in good agreement and proved suitable for single lap joints [5]. In the case of more complex geometrical configurations, it is impossible to analytically describe the stress field within the adhesive and it is then necessary to use a finite element analysis (FEA) This study covers the behavior of adhesively bonded joints as predicted by FEA. Abaqus has been utilized to investigate the stress distribution along the adhesive layer of three different joint types while under static tensile loading the analysis focused on the central stress of the adhesive joint along line [A-B] in the lap joints modelled as shown in Fig. 4a on simple-lap joint (SLJ), single-step-lap joint (joint1) and single-step-lap joint with release at the edges of the bond area (joint2) as shown in Fig. 4b.

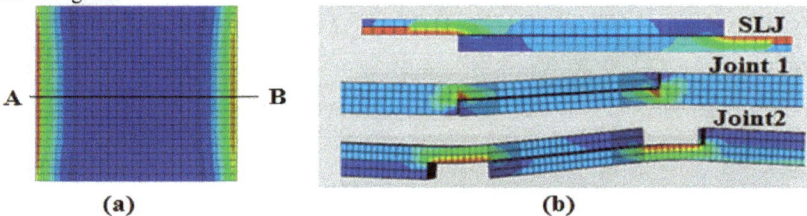

Fig. 4: (a) The Equivalent Von Mises stress distribution in the adhesive layer; (b) The stress concentration in the region of geometric discontinuity in different lap joints.

25

MECA SENS 2017   Materials Research Forum LLC
Materials Research Proceedings **4** (2018) 23-28   doi: http://dx.doi.org/10.21741/9781945291678-4

An adhesive joint is characterized by the presence of tensile stress at the ends of the joint, perpendicular to the plane of the adhesive. These peel stresses are caused by the bending of the substrates due to the loading eccentricity, inducing the presence of a bending moment in the joint. In this study, we undertook to investigate a stepped adhesive joint so as to minimize the effect of the bending moment. Fig. 5a shows the geometry of the stepped joint considered. Fig. 5b shows the most important stresses in the joint, Von Mises, S_{xx} (S11) and S_{yy} (S22); the stress concentration in the region of geometric discontinuity is evident.

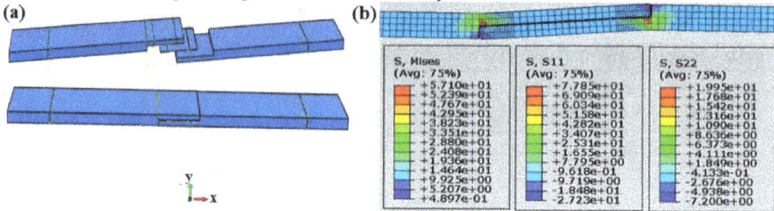

Fig. 5 :(a) Single-Step Lap Joint;(b)Variation of the stress state along the length of overlap joint.

We have noticed that instead of attenuating the stresses in the adhesive, the stepped adhesive joint as defined above does not improve the state of stress with respect to the single-overlap joint. One way to improve the behavior of the stepped adhesive joint would be to release the embedded region, that is, the junction between the overlapped portion and the remainder of the plate. In a stepped adhesive joint the overlapped part of the plate has no free space to flex under the effect of the bending moment. Free edges were allowed so as to release the overlapping portion of the plate, as shown in Fig. 6.

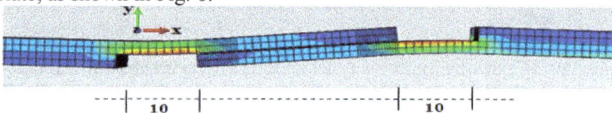

Fig. 6: Single-step joint with release 3D view (joint 2)

Comparison of the stress state in Joint1 and SLJ (single-lap joint). Figure 7 shows the Von Mises stresses, shear stress, peel stresses along the center line of the overlap in a single-lap joint (SLJ) of and the stepped joint (Joint1). The maximum Von Mises stress and shear stress for a stepped adhesive joint is slightly greater than that of the single-layer joint. In the same way as the Von Mises and shear stresses, the same remarks apply for the peeling stresses.

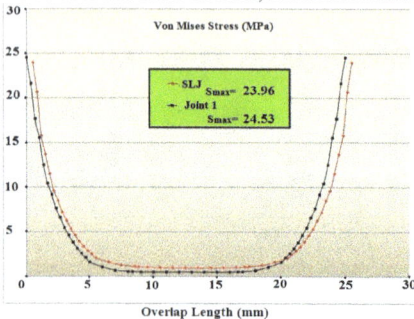

Fig. 7: Comparison of the adhesive Von Mises stress distributions, shear stresses τ_{xy} and the peel stresses σ_{yy} between SLJ and Joint1 along the overlap length (continue on next page).

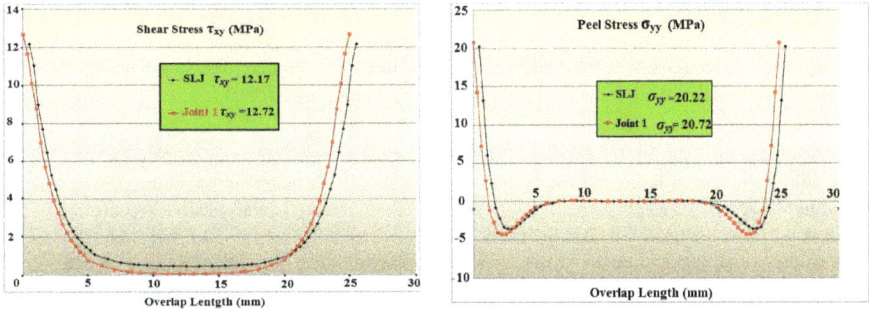

Fig. 7: (Continue from previous page) Comparison of the adhesive Von Mises stress distributions, shear stresses τ_{xy} and the peel stresses σ_{yy} between SLJ and Joint1 along the overlap length.

Comparison of the stress state in joint1, joint2 and single-lap joint.

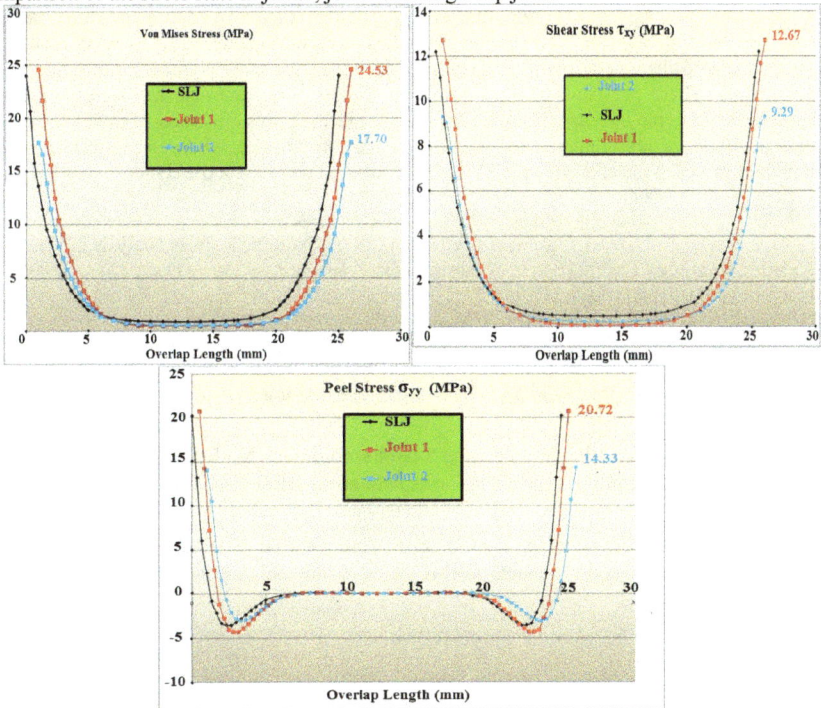

Fig. 8: Variation of the Von Mises stress, shear stress and peel stress distribution along the length of overlap.

Figure 8 shows that for the adhesive joint with release of the Von Mises stress have decreased by 28 % compared to that of the single-step joint (Joint1). For the released adhesive joint (Joint2) the peeling stress has decreased by 32 % compared to the single-step joint (Joint1) and the shear stress has decreased by 26 % compared with the single-step joint (Joint1).

Conclusions

In this study, the interfacial adhesive stress distribution of three different joint types, i.e. single-lap joint, single-step lap joint and single-step lap joint with edge release, subjected to tensile loading were investigated numerically by FEM and validated analytically. Accordingly, the following conclusions can be deduced. Good correlation was found between the FEM simulations and the analytical results. The development of a non-linear finite element model approach to simulate the different adhesive joints subjected to static tensile loading has given confidence in the results for bonded assemblies. Changing the geometry of the area in which the bonding process is performed, i.e. using one, two or three steps considering the edge release at the end of bond area has a profound impact on the stress concentrations forming at the adhesive joint and load carrying capacity of the joint. According to the results for the joints having the same bonding area, single-step lap joints with an edge release at bond area the stresses are decreased by about 28 % compared to other joints. The overlap area has a significant effect on the peel and shear stresses. This is due to the geometry of the overlap area which causes a secondary bending in the single lap joint. Designing steps for the adhesively bonded joints at the regions close to the edges of the overlap area decreased the peel stresses at the edges of the overlap area. These peel stresses are effective in initiating damage and this decrease played a significant role in the increase of the joint strength. For the joints, both shear (τ_{xy}) and peel (σ_{yy}) stress distributions are homogeneous along the width (bond line), while, the distributions have maximum values at the ends along the length and minimum values at the center.

References

[1] O. Sayman, A. Ozel, A. Pasinli and M. Ozen (2013), Nonlinear stress analysis in adhesively bonded single-lap joint, J. Adhes. Sci. Technol. 27 (21)(2013) 2304-2314. https://doi.org/10.1080/01694243.2013.773696

[2] O.Volkersen, Die Niektraftverteilung in Zugbeanspruchten mit Konstanten Laschenquerschnitten, Luftfahrtforschung 15 (1938) 41-68.

[3] M. Goland and E. Reissner: The stresses in cemented joints, J. Appl. Mech, ASME, New York, Vol 11, 1944.

[4] K. Madani, S. Touzain, X. Feaugas, S. Cohendouz and M. Ratwani, Experimental and numerical study of repair techniques for panels with geometrical discontinuities, Comput. Mater. Sci. 48 (2010) 83–93. https://doi.org/10.1016/j.commatsci.2009.12.005

[5] M. Khodja, A. Ramdoum, A. Bouakkaz, A. Amiri and W. Oudad, Etude et modélisation d'un joint à simple recouvrement, CNMI 3ème Conférence Nationale de Mécanique et d'Industrie. 3rd, (2014)

[6] Simulia, Dassault Systems. Abaqus software, http://www.3ds.com. Version 6.11, (2011)

Materials Research Forum LLC
doi: http://dx.doi.org/10.21741/9781945291678-5

Investigating Stresses Developed during Mechanical Forming of Steel through Finite Element Analysis

S.A. Akinlabi[1,a*], O.S. Fatoba[2,b] and E.T. Akinlabi[2,c]

[1]Department of Mechanical and Industrial Engineering Technology, University of Johannesburg, South Africa

[2]Department of Mechanical Engineering Science, University of Johannesburg, South Africa

[a]stephenakinlabi@gmail.com, [b]drfatobameni@gmail.com, [c]etakinlabi@uj.ac.za

Keywords: Finite Element Analysis, Stresses and Structural Integrity

Abstract. Stresses majorly affect the mechanical properties of materials. However, structural failures are often caused by the combined effect of residual stresses and applied stresses. It is practically impossible for a manufactured component to be entirely free of residual stresses because these stresses developed during the manufacturing process and certain amount remain in the component even after the process is completed. This study reports the findings of the investigation into the developed stresses during mechanical forming of the steel sheet. The result revealed that the Von Mises stresses developed, increases during the forming process. Also, the original tensile stresses in the material changed to compressive stresses along the inner radius as the punch strokes increases. Lastly, it was observed that the locked in stresses in the material after the process were tensile in nature and such are not beneficiary to the structural integrity of the manufactured component even though an average value of 0.057540 MPa was recorded for this study at the bend radius, distance away from the neutral plane.

Introduction

Finite element analysis has been a useful tool in research and engineering in particular for optimising and validating experimental procure and results. It has also been viewed as a tool deployed into design projects to save time and money through preliminary work using finite element analysis to establish the possible parameters and results. The experimental results are further validated with the finite element analysis results. Several types of research have deployed FEM to several manufacturing processes today. The demand in the manufacturing industry to design complex shapes to meet the new realities is the increase in the twenty-first century. The automotive industry, in particular, requires that the body shape of the cars be subjected to multiple and complicated loadings but also a strain rate that ranges between 0.1 s^{-1} and 100 s^{-1}. Metal sheet forming under this conditions are often characterised with shape and surface distortions, spring back, wrinkling and induced stresses. Finite element methods are a useful tool and approached to manage the defect by conducting finite element analysis to estimate the expected results [1-3].

Furthermore, it is known that the material properties of steel sheets are influenced by intermediate strain rates among other things [4]. At intermediate strain rates higher than a strain rate of tens per second, materials experience the effect of the inertia and the stress wave propagation, consequently greatly altering the material properties. Various techniques, such as the mechanical method [5] and the drop weight method [6], have been tried to measure the material properties at intermediate strain rates.

Residual stresses would unavoidably be induced in the sheet material in the most manufacturing process, but their magnitude would depend on whether the sheet is produced by cold or hot working conditions. Rossini *et al.* [7] reported that residual stresses originated from some sources, which are introduced during manufacturing or in-service loading. These stresses can be present even in the unprocessed raw material. Furthermore, they suggested that the origin of residual stresses could be classified according to differential plastic flow, differential cooling rates, and phase transformations with volume changes.

According to the Welding Institute (TWI) [8], these stresses as being caused by incompatible internal permanent strains that may be generated at every stage of the life cycle of a component; from the original production to the final disposal. Withers [9] on the other hand, considered residual stresses to be developed because of misfits (incompatibilities) between different regions of the material, sample or assembly. Flaman [10] classified residual stresses according to how it developed. Kannatey-Asibu [11] defined residual stresses as stresses that continue to exist in a material even when no external forces are acting on it. He believed that such stress could be produced in some ways, some of which are during welding, grinding, bending, forming, and heat treatment processes. The effect can be detrimental when it reduces the tolerance of the material to an externally applied force. In this paper, the authors' present an investigation into the stresses developed during mechanical forming of steel through Finite Element Analysis.

Mechanical Forming Process
The process of bending result in both tension and compression in the sheet [12], with the outer radius and the inner radius of the sheet, undergo tension and compression respectively. The schematic of a bending process is shown in Fig. 1. Some of the standard parameters in a bending process such as bend angle, neutral axis, bend allowance and set back are shown.

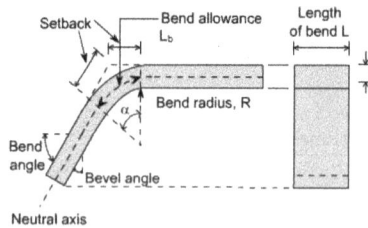

Fig. 1: Schematic of a bending Operation [13].

This phenomenon may be related to the bend allowance and bend deduction. However, it is important to state that due to the plastic deformation, there is residual stresses and strains after the forming process. These residual stresses consequently brings about elastic recovery in the material often called spring back which causes shape error in the final fabricated part.

Finite Element Analysis
Finite element analysis remain a tool that would continue to be relevant in all spheres of endeavour. This was applied to sheet forming process to evaluate the response and the mechanical properties of the material under loading. Marc software, 2015 version was employed for the analysis. The analysis is static but with elastic - plastic material characteristics.

Assumptions
The following assumptions were made:
- Analysis is static and assumed to be a simply supported;
- The hardening rule was based on Isotropic Model;
- The yield criterion was based on Von Mises;

MECA SENS 2017   Materials Research Forum LLC
Materials Research Proceedings **4** (2018) 29-34   doi: http://dx.doi.org/10.21741/9781945291678-5

- The Poisson ratio was assumed to be 0.3;
- The punch has initial contact with the sheet;
- A constant time stepping approach was employed with 0.01 s and 0.025 s for loading and unloading respectively;
- Convergence tolerance of 0.1 %;

Boundary Condition
- Constrain movements of sets of nodes along the X-axis at the contact of the punch and the plate;

The geometry of the steel was defined in Mentat Marc, a plane strain element was used, and the material properties and boundary conditions were set up. The schematic diagram of the mechanical forming setup showing both the deformable and rigid bodies and the resulting deformation at five-punch stroke are shown in Fig. 2.

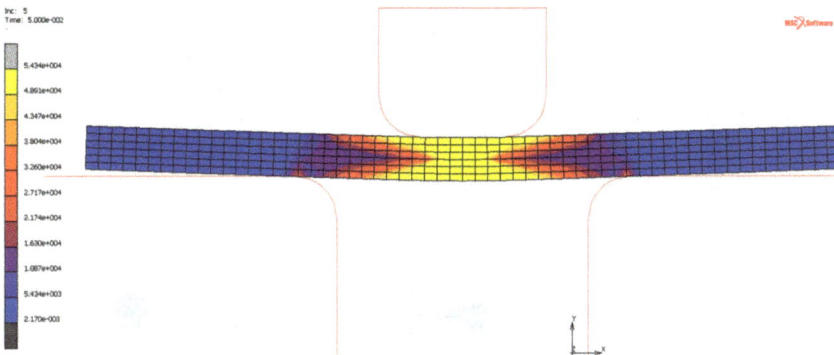

Fig. 2: Schematic of Mechanical Forming process setup and Stress Distribution at fifth stroke.

Result and Discussion
The finite element of the static analysis with elastic-plastic behaviour was conducted. In this study, the development of Von Mises stress was investigated and reported. The induced stresses into the sheet at the repeated punch stroke cycle of five were mainly tensile as displayed on the contour plot. It was also noted that the cloud of induced tensile stresses mainly concentrated around the contact area between the punch and the sheet, this was in agreement with the findings of both Rossini *et al*. [7] and Kannatey-Asibu [11].

The mechanical process of this nature in manufacturing is characterised by both tension and compression in the process, whereby consequently induce stresses some of which are naturally relieved during or at the end of the process while the rest remain within the manufactured or processed material. This phenomenon has been demonstrated through this study.

It was also noted that the bend angle increased from 9.8 degrees to 52.3 degrees at the fifth strokes, this is very significant and as such advised to be monitored during the manufacturing process.

The distribution of the sets of measured Von Mises stresses with increasing number of strokes is shown in Fig. 3. The plotted stress distributions show a progressive increase in both the inner and outside radius as the punch stroke cycle increases.

MECA SENS 2017   Materials Research Forum LLC
Materials Research Proceedings **4** (2018) 29-34   doi: http://dx.doi.org/10.21741/9781945291678-5

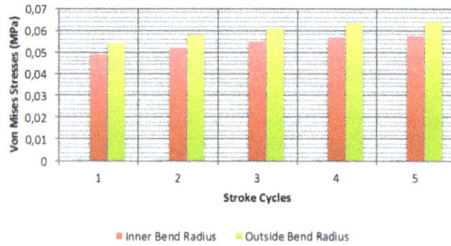

Fig. 3: Distribution of the Von Mises Stress.

Further analysis of the developed stresses during the mechanical forming process revealed the evolutional changes that characterised the forming process, which is critical to establish the stroke threshold that will give tolerable stresses considering the structural integrity that may be desired for the formed component. Four developed Von Mises Stresses were measured but only two at stroke cycles two and five were shown in Figs. 4 and 5. The process of sheet forming employs tensile force in the plane of the transverse direction of the sheet this consequently induces multi-stresses condition in the sample causing the bending of the sheet. This process entails both tension and compression in the sheet due to the punch impact on the sheet.

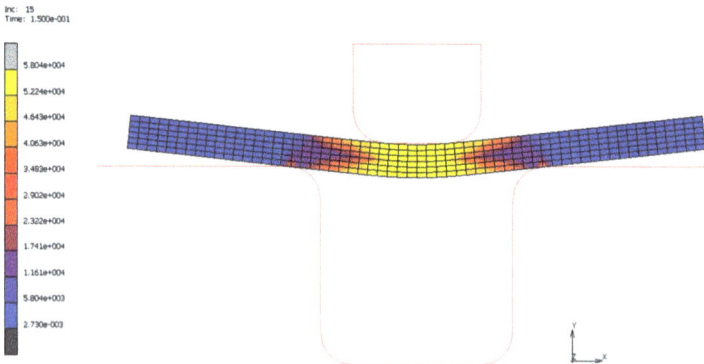

Fig. 4 : Equivalent Von Mises Stresses measured at stroke cycles of two.

It is observed that the impact of the stroke cycles in the forming of the sheet was significant. The first thing worth noting was the tensile nature of the developed stress along both the inner and outside bend radius but less compressive along the inner bend radius. However, the trend observed in all the samples was that the developed stresses become less tensile and more compressive along the neutral axis of the sheet as the bending process progresses, even though experimentally stresses at the neutral axis tend towards zero. This points to the importance of the neutral axis in sheet forming operation. The neutral axis is the plane separating the inner bend radius from the outer bend radius. As the number of the cycle stroke increases, the dimensions of the thickness of the neutral axis increases to as high as 1.5 mm, such an increase was observed to be very significant.

Hence, effective sheet forming operation controls and ensures the stress locked around the neutral axis is within tolerable values otherwise it can be catastrophic. This further confirms the

MECA SENS 2017   Materials Research Forum LLC
Materials Research Proceedings **4** (2018) 29-34   doi: http://dx.doi.org/10.21741/9781945291678-5

significant role of the punch stroke during a sheet forming operations. Another important observation worthy of highlighting is the formation of cup-like configuration around the bend area. This cup-like shape is a mirror of the punch geometry and as such this will be an excellent tool to control the developed stresses through the design of the punch geometry and material.

The relaxation during the first unloading during the forming process is shown in Fig. 6. During the unloading of the punch at the end of the complete cycle, even though the nature of the stress remains tensile and compressive, a significant stress relaxation was observed. The magnitude of the stress around the neutral axis during the last loading stroke was 0.057540 MPa while at the first unloading stroke, the magnitude of the stress along the neutral axis was 0.031650 MPa. This relaxation phenomenon is often called elastic recovery.

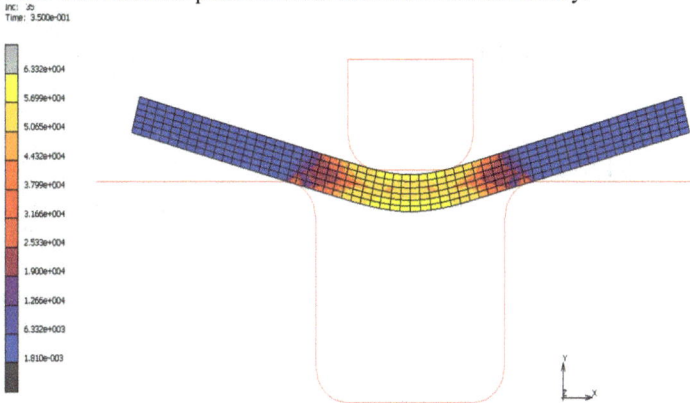

Fig. 5: Equivalent Von Mises Stresses measured at stroke cycles of five.

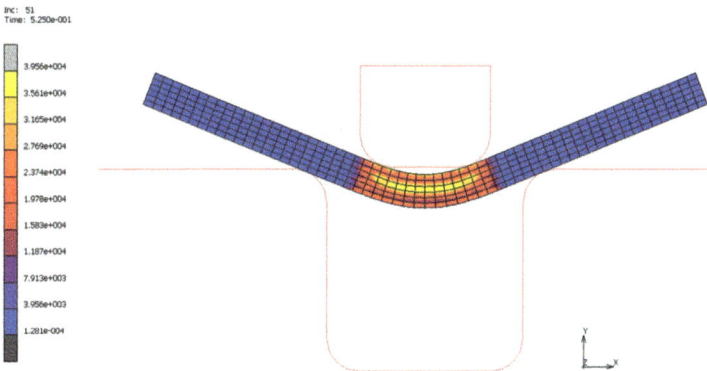

Fig. 6: Relaxation at the first unloading during the forming Process.

Conclusion

The finite element analysis of the mechanical forming of sheet steel was conducted and completed. The following conclusions can be drawn:

- Tensile and compressive stresses are developed during the mechanical forming process, and it increases as the punch stroke increases.

- The stresses along the inner and outside radius are tensile while stresses close and around the neutral axis are more compressive.
- The stress concentration close to the neutral axis increases as the punch strokes increases.
- The stress distributions form a cup-like configuration, which is a reflection of the geometry of the punch.
- An elastic recovery of 0.025890 MPa developed at the unloading of the punch.

References

[1] G. Joo, H. Huh and M.K. Choi, Tension/Compression hardening behaviors of auto-body Steel Sheets at Intermediate strain rates, Int. J. Mech. Sci. 108-109 (2016) 174-187. https://doi.org/10.1016/j.ijmecsci.2016.01.035

[2] H. Huh, J. H. Lim and S. H. Park, High speed tensile test of steel sheets for the stress - strain curve at the intermediate strain rate, Int. J. Mech. Sci. 2009 10(2) 195–204. https://doi.org/10.1007/s12239-009-0023-3

[3] M.A. Meyer, Dynamic behavior of materials, John Wiley & Sons, N.Y, U.S.A, 1994. https://doi.org/10.1002/9780470172278

[4] D.R. Ambur, C.B. Prasad and W.A. Waters Jr., A dropped-weight apparatus for Low Speed impact testing of composite structures, Exp. Mech. 35 (1995) 77–82. https://doi.org/10.1007/BF02325839

[5] N.S. Rossini, M. Dassisti, K.Y. Benyounis and A.G. Olabi. Methods of measuring residual stresses in Samples, Material and Design, 35 (2012) 572-588. https://doi.org/10.1016/j.matdes.2011.08.022

[6] TWI Knowledge Summary, 2015, www.twico.uk/j32/protected/band_3/ksrh1001.html [Accessed February 2017]

[7] P.J. Withers, Residual stress and its role in failure, Report of progress in Physics 70 (2007) 2211-2264. https://doi.org/10.1088/0034-4885/70/12/R04

[8] M.T. Flaman, Investigation of ultra-high speed drilling for residual stress measurements by the Centre-hole method, Exp. Mech. 22 (1) (1982) 26-30. https://doi.org/10.1007/BF02325700

[9] E. Kannatey-Asibu, Principle of Laser Materials Processing, John Wiley & sons, Inc., N.Y., 2009. https://doi.org/10.1002/9780470459300

[10] Sheet Metal Processes, http://nptel.ac.in/courses/112107144/6 [Accessed March 2017].

[11] G.H. Bae and H. Huh, Tension/compression test of auto-body steel sheets with the variation of the pre-strain and the strain rate, In: Proc. 5th Int. Conf. on Computational Methods and Experiments in Materials Characterisation, Kos, Greece. 2011. https://doi.org/10.2495/MC110191

[12] S. Choudhry and J.K. Lee, Dynamic plane-strain finite element simulation of industrial sheet-metal forming processes, Int. J. Mech. Sci. 36 (1994) 189–207. https://doi.org/10.1016/0020-7403(94)90069-8

[13] H. Tze-Chi and C. Chan-Hung, A finite element analysis of sheet metal forming processes, J. Mater. Process. Technol. 54 (1995) 70-75. https://doi.org/10.1016/0924-0136(95)01922-7

Fatigue, Creep & Plasticity

MECA SENS 2017   Materials Research Forum LLC
Materials Research Proceedings **4** (2018) 37-42   doi: http://dx.doi.org/10.21741/9781945291678-6

Dislocation Density of Oxygen Free Copper with Compressive Strain Applied at High Temperature

M. Sano[1,a*], S. Takahashi[1,b], A. Watanabe[1,c], A. Shiro[2,d] and T. Shobu[3,e]

[1]JASRI, 1-1-1 Kouto Sayo-cho, Sayo-gun, Hyogo 679-5198, Japan

[2]QST Quantum Beam Research Directorate 1-1-1 Kouto Sayo-cho, Sayo-gun, Hyogo 679-5148, Japan

[3]JAEA Materials Sciences Research Center, 1-1-1 Kouto Sayo-cho, Sayo-gun, Hyogo 679-5148, Japan

[a]musano@spring8.or.jp, [b]takahasi@spring8.or.jp, [c]watanaba@spring8.or.jp, [d]shiro.ayumi@qst.go.jp, [e]shobu@spring8.or.jp

Keywords: Dislocation Density, OFC, Plastic Strain, Line Profile Analysis

Abstract. Dislocation densities of oxygen-free copper (OFC) with compressive strain applied at high temperatures were examined by X-ray line profile analyses with synchrotron radiation. To evaluate the dislocation density, we applied the modified Williamson-Hall and modified Warren-Averbach methods. The dislocation densities of OFC with compressive strain ranging from 0.9 – 3.8 % were on the order of $1.2 \times 10^{13} - 4.2 \times 10^{14}\,\mathrm{m^{-2}}$.

Introduction

Oxygen-free copper (OFC) is one of the most popular materials for use in high heat load components that are essential components of many accelerator facilities. At the SPring-8 front-end section, OFC has been incorporated into high heat load components such as photon absorbers and masks in all bending magnet and some undulator beamlines that are subjected to synchrotron radiation with a relatively low power density. Although these components have operated without any thermo-mechanical problems for more than fifteen years, the thermal limitations of OFC should be investigated to deal with increased heat loads in the future, considering the SPring-8 upgrade plan. Recently, we investigated the fatigue phenomenon of OFC and performed residual strain measurements on the OFC samples by using synchrotron radiation [1, 2]. In addition to these investigations, we have recently been examining the relationship between the plastic strain and the dislocation density for the high heat load materials [3, 4].

The purpose of this study is to examine the relationship between the plastic strain and the dislocation density for OFC, as these two parameters are generally correlated with each other. Recently, X-ray line profile analysis has emerged as one of the most powerful methods for the nondestructive investigation of dislocation structures in plastically deformed materials [5]. We examined the dislocation density of OFC samples with compressive strain applied at high temperature. Modified Williamson-Hall and modified Warren-Averbach methods were applied to estimate the dislocation density.

Experimental

Standard OFC samples, which were of grade C1011 (99.99 % Cu), had known values of compressive plastic strains. The initial configuration of the samples was a cylinder with a height and diameter of 15 mm. These samples were manufactured with compressive plastic strains ranging from 0.9 – 3.8 % applied at approximately 300°C at an approximate strain rate $6.7 \times 10^{-5}\,\mathrm{s^{-1}}$.

The central volumes of the samples, with thicknesses of 2 mm, were cut by electrical discharge machining after the compression stage.

The diffraction experiments were performed at beamline BL02B1 of SPring-8 with a monochromatic beam of 72.3 keV (λ =1.72×10^{-2} nm) and a Pilatus3 X CdTe 300K 2D detector. Table 1 shows the optical configuration and experimental conditions used for the strain measurements. The strain scanning method with oscillation was used because of the large grain size of the OFC samples. The oscillation length was ±3 mm in the vertical direction. The measurements were performed at the center of the samples. Fig. 2 shows representative diffraction profiles for the Cu (200), (311), (400), (331), and (422) reflections of the OFC samples under a compressive plastic strain of 2.6 %.

Table 1. Experimental conditions.

Beam line	SPring-8/BL02B1
Measurement method	Transmission-type strain scanning method
Energy [keV]	72.3
Monochromatic crystal	Si(311)
Diffraction angle (2θ) [°]	3.5 – 13.5
Camera length [mm]	1268
Slit size (Width × Height) [mm^2]	Divergent Slit 1: 2 × 0.2

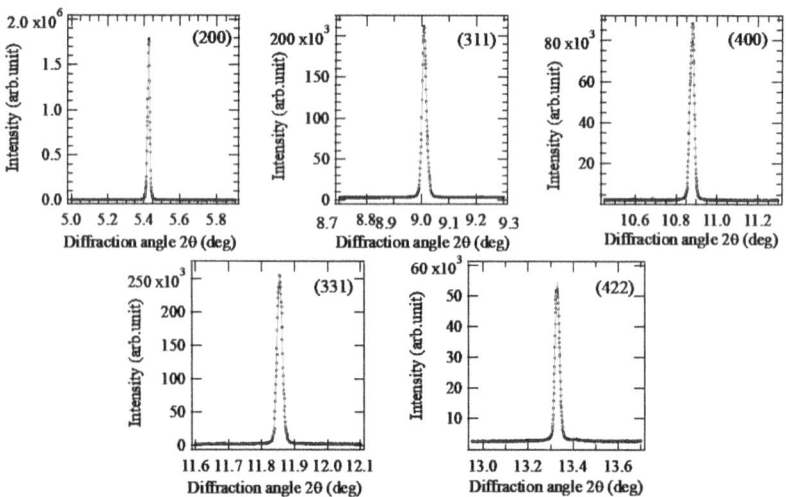

Fig. 1: Cu (200), (311), (400), (331), and (422) diffraction profiles of the OFC sample under a compressive plastic strain of 2.6 %. The symbols represent experimental data and the solid line corresponds to the fitting of a pseudo-Voigt function with a linear background.

Line Profile Analysis

As shown in Fig. 1, a pseudo-Voigt function with a linear background was applied to the profiles as a fitting function. In this study, it was assumed that instrumental line broadening was negligible, as the instrumental line broadening of the beamline was expected to be less than 0.002°, according to a previous study [5].

The dislocation density was evaluated using the modified Williamson-Hall and modified Warren–Averbach methods, which are based on the FWHM value and the Fourier coefficient of the diffraction profile [5, 6]. The FWHM values the Fourier coefficients were obtained from the fitting functions. Assuming that the dislocations mainly contribute to the line broadening caused by strain, the modified Williamson-Hall method can be expressed by the following equation:

$$\Delta K \cong 0.9/D + \sqrt{\pi M^2 b^2/2} \sqrt{\rho} K\sqrt{\bar{C}} + O(K^2\bar{C}). \tag{1}$$

where, $K = 2\sin\theta/\lambda$, ΔK is the FWHM, D is the average particle size, M is a constant, b is the absolute value of the Burgers vector, ρ is the dislocation density, $\bar{C}$ is the average contrast factor of the dislocations, and O indicates higher order terms in $K\sqrt{\bar{C}}$. In a cubic crystal system, the average contrast factor can be described as follows:

$$\bar{C} = \bar{C}_{h00}(1 - qH^2). \tag{2}$$

where $\bar{C}_{h00}$ is the average contrast factor corresponding to the ($h00$) reflection, q is a constant, and $H^2 = (h^2k^2 + h^2l^2 + k^2l^2)/(h^2 + k^2 + l^2)$. Inserting Eq. 2 into Eq. 1 yields

$$((\Delta K)^2 - \alpha)/K^2 \cong \beta\bar{C}_{h00}(1 - qH^2). \tag{3}$$

where $\alpha = (0.9/D)^2$ and $\beta = \pi M^2 b^2 \rho/2$. From a linear regression of the left-hand side of Eq. 3 and H^2, the parameter q can be determined. As shown in Fig. 2, the q values were obtained from the modified Williamson-Hall method for the compressive plastic strains of 0.9 % and 2.6 %. The average contrast factor, $\bar{C}_{h00}$, of copper 0.304 [6] was used in the analysis.

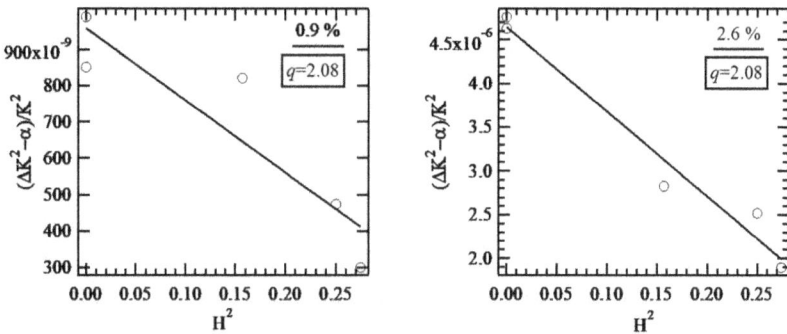

Fig. 2: Relationship between $((\Delta K)^2 - \alpha)/K^2$ and H^2 under compressive plastic strains of 0.9 % and 2.6 %. The solid line shows the fit of the data to Eq. 3.

MECA SENS 2017   Materials Research Forum LLC
Materials Research Proceedings 4 (2018) 37-42   doi: http://dx.doi.org/10.21741/9781945291678-6

The dislocation density can be determined from the Fourier coefficients by applying the modified Warren–Averbach method:

$$\ln A(L) \cong \ln A^s(L) - (\pi b^2/2)\rho L^2 \ln(R_e/L)(K^2\bar{C}) + O(K^2\bar{C})^2. \tag{4}$$

where $A(L)$ is the real part of the cosine Fourier coefficient of the diffraction profile, A^s is the size Fourier coefficient, L is the Fourier length, R_e is the effective outer cut off radius of the dislocation, and O represents higher-order terms in $K^2\bar{C}$. By fitting the left-hand side of Eq. 4 as a quadratic function of $K^2\bar{C}$, $A^s(L)$ and the slope $(L) = (\pi b^2/2)\rho L^2 \ln(R_e/L)$ can be obtained. As shown in Fig. 3, using the $A(L)$, the modified Warren–Averbach method was applied to the compressive plastic strains of 0.9 % and 2.6 %. The value of the slope, $X(L)$, for each L value could be determined from the fitting of Eq. 4.

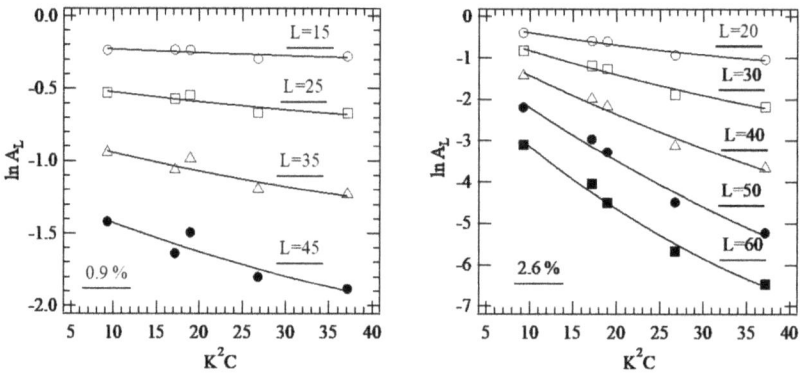

Fig. 3: The relationship between $\ln A_L$ and K^2C for each L value for compressive plastic strains of 0.9 % and 2.6 %. The solid line shows the fit of the data to Eq. 4.

The slope $X(L)$ can then be evaluated according to the following equation:

$$X(L)/L^2 = \rho(\pi b^2/2)(\ln R_e - \ln L). \tag{5}$$

From the linear regression of $X(L)/L^2$ and $\ln L$, the dislocation density, ρ, can be obtained. The dislocation densities were evaluated from a linear regression described by Eq. 5, as shown in Fig. 4.

Results and Discussion

Figure 5 shows changes in the q values with error bars and the dislocation character with theoretical values for each dislocation type, as a function of compressive strain. While half edge – half screw dislocations were predominant in the range between 0.9 and 2.6 %, the character was pure screw dislocation for 3.8 %. Figure 6 shows the dislocation density values with error bars as a function of compressive strain. These error bars were obtained from only the fitting by Eq. 5. The dislocation density rapidly increased from 1.2×10^{13} to 3.6×10^{14} m^{-2} when the compressive strain increased from 0.9 to 2.6 %. On the other hand, the dislocation densities only increased gradually to 4.2×10^{14} m^{-2}, when the compressive strain was changed from 2.6 to 3.8 %.

The dislocation densities of the OFC samples with compressive strain applied at room temperature were 5.1×10^{14} and 9.2×10^{14} m^{-2} for compressive plastic strains of 1 % and 4 %, respectively [4]. The values were smaller at 300 °C than at room temperature within the measured strain range, because it was considered that the mobility of dislocations was higher at high temperature.

As seen in Fig. 5, the error bars of q values were large especially in the case of 0.9 %. Improvements will be needed in both the experiment and analysis to obtain more precise values for the dislocation density, since it depends on the dislocation contrast factor.

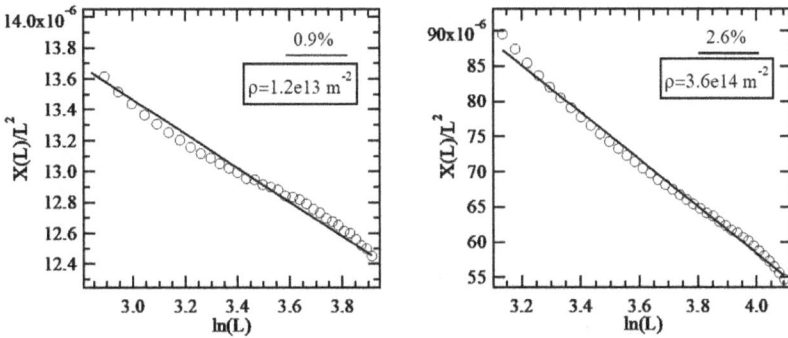

Fig. 4: Relationship between $X(L)/L^2$ and $\ln L$ for compressive plastic strains of 0.9 % and 2.6 %. The solid lines shows the fits of the data to Eq. 5.

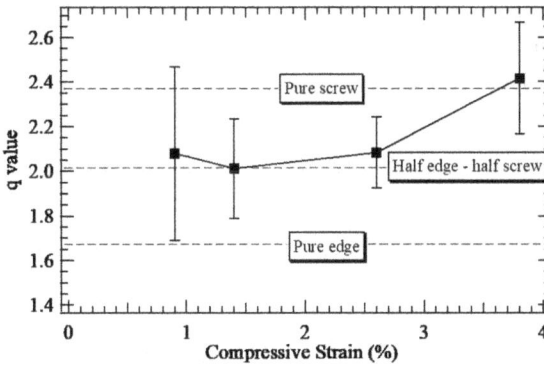

Figure 5: Relationship between the compressive strains and q. The horizontal dotted lines show the theoretical values for each dislocation type.

Summary
In this study the dislocation densities of OFC samples with compressive strains applied at high temperature were estimated by applying the modified Williamson-Hall and modified Warren-Averbach methods. The dislocation densities and the characters of the dislocations were

obtained over a strain range of 0.9 – 3.8 %. The dislocation densities of the OFC samples ranged from 1.2×10^{13} to 4.2×10^{14} m^{-2} within the specified strain range.

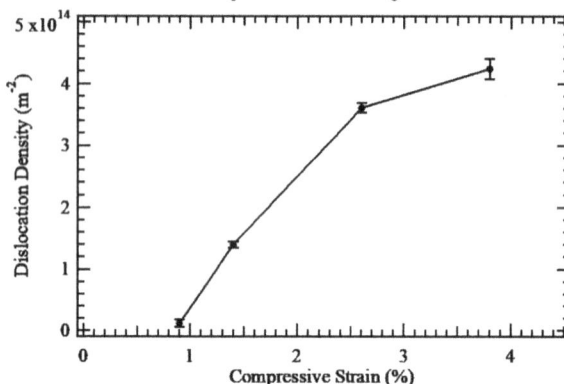

Fig. 6: Relationship between the compressive strain and dislocation density.

Acknowledgements

The synchrotron radiation experiments were performed at the SPring-8 with the approval of the Japan Synchrotron Radiation Research Institute (JASRI) (Proposal No. 2016A1335). Use of the modified Williamson-Hall and Warren-Averbach methods developed by Prof. Ungar's work (see references) is acknowledged.

References

[1] S. Takahashi, M. Sano, A. Watanabe and H. Kitamura, Prediction of fatigue life of high-heat-load components made of oxygen-free copper by comparing with Glidcop, J. Synchrotron Rad. 20 (2013) 67-73. https://doi.org/10.1107/S0909049512041192

[2] M. Sano, S. Takahashi, A. Watanabe, A. Shiro and T. Shobu, Residual strain of OFC using synchrotron radiation, Mat. Sci. For. 777 (2014) 225-259. https://doi.org/10.4028/www.scientific.net/MSF.777.255

[3] M. Sano, S. Takahashi, A. Watanabe, A. Shiro and T. Shobu, Dislocation Density of GlidCop with Compressive Strain applied at High Temperature, Mat. Res. Proc. 2 (2016) 609-614.

[4] M. Sano, S. Takahashi, A. Watanabe, A. Shiro and T. Shobu, Dislocation Density of Plastically Deformed Oxygen-Free Copper, Mat. Sci. For. 905 (2017) 60-65. https://doi.org/10.4028/www.scientific.net/MSF.905.60

[5] T. Ungar and A. Borbely, The effect of dislocation contrast on X-ray line broadening: A new approach to line profile analysis, Appl. Phys. Lett. 69 (1996) 3173-3175. https://doi.org/10.1063/1.117951

[6] T. Ungar, I. Dragomir, A. Revesz and A. Borbely, The contrast factors of dislocations in cubic crystals: the dislocation model of strain anisotropy in practice, J. Appl. Cryst. 32 (1999) 992-1002. https://doi.org/10.1107/S0021889899009334

[7] Y. Noda, Current Status of Crystal Structure Analysis BL02B1 Experimental Station, SPring-8 INFORMATION, Volume 02, No.5 (1997) 17-23.

Mechanical Methods vs. Diffraction

MECA SENS 2017   Materials Research Forum LLC
Materials Research Proceedings **4** (2018) 45-50   doi: http://dx.doi.org/10.21741/9781945291678-7

Evaluation of Residual Stresses Introduced by Laser Shock Peening in Steel using Different Measurement Techniques

D. Glaser[1,2,a*], M. Newby[3,b], C. Polese[2,4,c], L. Berthe[5,d], A.M. Venter[4,6,e],
D. Marais[6,f], J.P. Nobre[2,g], G. Styger[7,h], S. Paddea[8,i] and S.N. van Staden[2,j]

[1]CSIR National Laser Centre, Brummeria, Pretoria, South Africa

[2]University of the Witwatersrand, Johannesburg, South Africa

[3]Eskom Holdings SOC Ltd, Johannesburg, South Africa

[4]DST-NRF Centre of Excellence in Strong Materials, University of the Witwatersrand, Johannesburg, 2000, South Africa

[5]Laboratoire PIMM (ENSAM, CNRS, CNAM, Hesam Université), Paris, France

[6]South African Nuclear Energy Corporation (Necsa) SOC Limited, Pretoria, 0001, South Africa

[7]University of Johannesburg, Auckland Park, Johannesburg, South Africa

[8] The Open University, Milton Keynes, United Kingdom

[a]DGlaser@csir.co.za, [b]Mark.Newby@eskom.co.za, [c]Claudia.Polese@wits.ac.za,
[d]Laurent.Berthe@ensam.eu, [e]Andrew.Venter@necsa.co.za, [f]Deon.Marais@necsa.co.za,
[g]JoaoPaulo.Nobre@wits.ac.za, [h]GaryStyger@gmail.com, [i]S.Paddea@open.ac.uk,
[j]Sean.vanStaden2@students.wits.ac.za

Keywords: Laser Shock Peening, Residual Stress, Synchrotron Diffraction, Incremental-Hole Drilling, Neutron Diffraction, Contour Stress Measurement

Abstract. The development of a residual stress engineering technology such as laser shock peening (LSP) requires evaluation of the process by quantification of the desired effect. Applications of LSP for turbine blade integrity enhancement due to expected deeper compressive residual stresses with lower surface roughness compared to conventional shot peening (SP), have resulted in the analysis of LSP on 12CrNiMoV steel samples. The investigation compares different residual stress measurement techniques such as energy dispersive synchrotron X-ray diffraction (SXRD), laboratory X-ray diffraction (XRD) with sequential electro-polishing, neutron diffraction (ND), incremental-hole drilling (IHD), and the contour method (CM). This study highlights the benefits and opportunities of using complimentary residual stress measurement techniques in order to gain insight into the residual stresses within a material.

Introduction

Laser shock peening (LSP) is a residual stress engineering technology specifically used to introduce beneficial compressive residual stresses into critical components to enhance fatigue and/or stress corrosion cracking (SCC) performance [1]. The current research has been conducted within a program focused toward development of LSP technology for applications of low pressure (LP) steam turbine blades which are typically susceptible to fatigue and SCC specifically in the blade attachment region as depicted in Fig. 1 [2]. Conventionally the highly stressed fir-tree attachment locations are mechanically shot peened (SP), however the laser-based LSP technology offers various benefits such as deeper levels of compressive residual stresses with improved surface roughness, as well as potentially being more suitable for reliable application to complex 3D surfaces. An integral aspect of the development of a residual stress

MECA SENS 2017   Materials Research Forum LLC
Materials Research Proceedings **4** (2018) 45-50   doi: http://dx.doi.org/10.21741/9781945291678-7

engineering technology such as LSP is quantification of the residual stresses introduced. This study explores the attributes of residual stress measurements conducted using synchrotron X-ray diffraction (SXRD), laboratory X-ray diffraction (XRD) in conjunction with electro-polishing, fine incremental-hole drilling (IHD), neutron diffraction (ND) and the contour method (CM).

Methodology

Sample generation: Samples were extracted from an ex-service LP steam turbine blade by removing slices from the fir-tree attachment region as depicted in Fig. 1. A stress relieving cycle of 660°C for 20 minutes was performed on the coupons. The samples were wire EDM cut to dimensions of 20 x 20 x 15 mm^3 and the surfaces ground. Electro-polishing was used to remove the surface grinding effects and hardness checked to ensure that the stress relieving did not alter the mechanical properties. Laboratory XRD measurements were performed on each sample before and after LSP processing to evaluate repeatability of the sample preparation and LSP processing. The work presented in this paper only considers the stresses in the y-direction as per the schematic in Fig. 1. Stresses are computed from the measured strains by utilizing a Young's modulus of 204 GPa and Poisson's ratio of 0.3. In the case of the XRD analyses the X-ray elastic constants were experimentally determined. The uncertainties associated with the diffraction based measurements of SXRD, and ND was typically less than ±10 MPa and XRD less than ±15 MPa, however these have been omitted from the presented results for purposes of clarity.

LSP processing: Initial proof-of-concept LSP processing was performed at the CSIR National Laser Centre in order to determine approximate LSP parameter combinations. Final processing was performed at the PIMM Laboratory (ENSAM-CNRS-CNAM) due to enhanced beam quality and an additional study that included parameter investigations, such as spot size effect, that require a high energy laser source. A Thales GAIA laser operating at a wavelength of 532 nm was used with sample immersion in a water tank. A black PVC tape (around 100 μm thick with a 30 μm adhesive) was employed as a sacrificial thermo-protective overlay, and the laser spot overlap was kept constant at 21.5 %. Using XRD as a screening analysis, an appropriate power intensity of 5 GW/cm^2 with a 2 mm spot size was identified as optimal parameters. The LSP processing was performed on the 20 x 20 mm^2 sample face with a 10 x 10 mm^2 LSP patch as depicted in Fig. 1 (except for a sample with an 18 x 18 mm^2 LSP patch specifically for neutron diffraction measurements). Only one application layer of LSP was used on the samples.

| Low Pressure Steam turbine rotor | turbine blade | Fir tree attachment location | LSP Coupon with coating after processing | LSP Coupon after removal of coating | |

Fig. 1: Photographs of a steam turbine rotor illustrating the location of samples within the LP blade and a schematic of the LSP processed coupons.

Laboratory X-ray diffraction (XRD): The residual stress analysis was performed using conventional laboratory XRD measurements with a Proto iXRD (Proto Manufacturing Inc., Taylor, Michigan USA) instrument. A Cr-Kα X-ray source with a wavelength of 2.291 Å was used in conjunction with a round 1.0 mm aperture. Reflections from the (211) peak for the steel were used, where the Bragg angle was located at 156.31°. The measurements were performed as per the sin^{2}ψ technique, whereby 20 exposures of 1 second each were used for 7 ψ measurement angles per strain measurement. Measurements were performed at 0, 45, and 90 degrees in order to obtain the principal stresses. Stresses were calculated using the X-ray elastic constants

MECA SENS 2017   Materials Research Forum LLC
Materials Research Proceedings **4** (2018) 45-50   doi: http://dx.doi.org/10.21741/9781945291678-7

measured using a 4-point bend apparatus as per the ASTM 146-91 designation which rendered $-S_1 = 1.115 \times 10^{-6}$ MPa^{-1} and $1/2S_2 = 5.247 \times 10^{-6}$ MPa^{-1}. Sequential layer removal and XRD measurements were performed in order to obtain the through-thickness stress variation. A Struers Lectro-Pol 5 was used to perform electro-polishing to remove around 15 to 100 microns of material per step, with finer increments used nearer to the surface. A correction factor proposed by Moore and Evans incorporated in the Proto software has been applied to the measured stresses in order to correct for the material removal [3].

Synchrotron X-ray diffraction (SXRD): The SXRD measurements were conducted at the ID15A beamline (experiment ME1440) at the ESRF facility in Grenoble, France. Energy-dispersive measurements were performed with 300 keV X-rays which allowed for transmission through the 20 mm dimension of the samples in order to provide strain measurements in the y-direction as indicated in the Fig. 1 schematic. The beam dimensions were set to 50 microns (orientated through the thickness) by 100 microns, and a diffracting angle of around 3° results in a gauge volume elongation to around 1.9 mm along the sample surface. A GSAS Pawley analysis was used in to determine the lattice parameter d_0 accounting for multiple peaks for the BCC material. A small pillar of the material (2 x 2 x 10 mm^3) was used to determine the stress-free lattice parameter.

Neutron diffraction (ND): The ND measurements were performed at the neutron strain scanner instrument MPISI located at the SAFARI-1 research reactor of Necsa (South African Nuclear Energy Corporation). Using a monochromatic wavelength of 1.67 Å the Fe (211) reflection manifests at a diffraction angle (2θ) of ~90°, which in conjunction with beam limiting apertures, defined an elongated cuboid gauge volume of 0.3 x 0.3 x 17 mm^3. To account for the d_0 value, a bi-axial stress condition was assumed in which the normal stress component is zero [6]. From this approach the d_0 value was determined at each measurement position and used for point-by-point strain calculation. Measurements were taken at intervals of 0.15 mm, starting at 0.25 mm from the peened surface to avoid partial filling of the gauge volume.

Incremental-hole drilling (IHD): The IHD has been performed using a SINT MTS3000 instrument which uses a high speed inverted cone carbide end mill. Type A strain gauges (CEA-13-062UL-120) have been used which have a nominal hole diameter of 2 mm, and stresses are reported to 50 % of the measured hole diameter of 1.75 mm. The drilling was performed with 60 increments of 0.02 mm each, and the stresses calculated as per the ASTM E837-13 EXT formulation for non-uniform residual stresses. The data was computed using the Eval Premium software where measured strains are interpolated with 20 linear steps. Since the IHD technique is considered a semi-destructive technique whereby execution of the tests is known to affect the reliability of the measurements [4], at least two repeatability trials have been conducted.

Contour method (CM): The CM stress analysis was performed across the sample centre in the x-direction as per the schematic in Fig. 1. The wire electric discharge machine (EDM) cutting used a 0.1 mm diameter uncoated brass wire, and a sacrificial material was bonded to the top surface to improve the near surface data quality. The cut surface profile was measured using a laser probe with a resolution of 25 x 25 x 0.15 μm^3. Data smoothing was implemented using an enhanced polynomial fitting routine which used a 3 mm smoothing mask. The residual stresses were computed by finite element analysis (FEA) which used a mesh resolution of 0.2 mm.

Results and Discussion
The repeatability between samples was first evaluated by XRD with line scans in the x-direction as indicated in Fig. 1 with 0.5 mm step increments and a 1 mm aperture. These results are provided in Fig. 2a which shows the stresses in the y-direction. The results in Fig. 2 refer to various samples prepared in an identical manner indicated by a letter in the legend of each graph.

MECA SENS 2017   Materials Research Forum LLC
Materials Research Proceedings **4** (2018) 45-50   doi: http://dx.doi.org/10.21741/9781945291678-7

The stresses vary in a range between -480 and -580 MPa over the 10 mm LSP region, where the average for each sample are more consistent with a variation from -514 to -543 MPa. The waviness or oscillatory nature of each XRD surface stress profile is expected to be due to residual stress variations between each LSP spot impact. Each LSP spot diameter is 2.0 mm with a 1.57 mm step distance between each spot. Subsequently, a XRD line scan has been performed with smaller apertures of 0.5 mm and 0.2 mm, and more distinctive variations are observed with a period corresponding to the spot overlap. Since the mean effects are consistent, the samples processing is considered repeatable. The variations of residual stress are therefore not spurious measurement phenomena, but rather physical feature due to the LSP spot overlap.

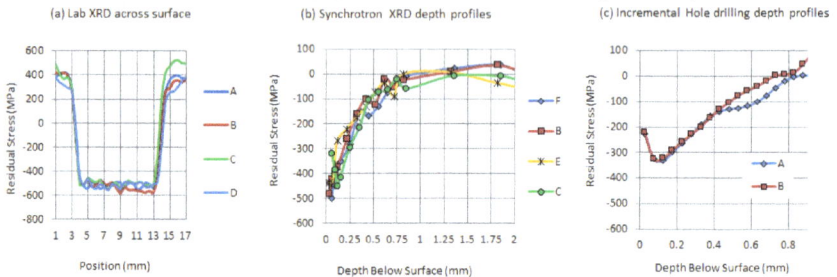

Fig. 2: Residual stress results of the LSP sample from: (a) XRD profiles across the LSP surface, (b) SXRD depth profiles below the LSP region and (c) incremental hole drilling depth profiles in the y- direction.

Comparisons of SXRD results from identical samples are shown in Fig. 2b. Notwithstanding the different SXRD measurements revealing similar trends, there are subtle variations between each measured profile. Previous investigations [2] on the same material processed by shot peening and evaluated on the same SXRD beamline reveal a smoother stress profile. This may suggest that the variations are not due to grain size effects, but rather potentially reflect variations within the LSP processed region. The variations between repeat samples suggest that the stress field is not as uniform as may be expected [5], which is supported by observation of the oscillation found with surface XRD measurements. If the LSP induced stress field varies according to the spot overlap, then the "needle like" gauge volume of the SXRD measurements $(0.05 \times 0.1 \times 1.9 \text{ mm}^3)$ may be sampling slightly different areas under the LSP pattern due to minor offsets during LSP processing or measurement.

The results from the two samples for IHD are shown in Fig. 2c. There appears to be good agreement between to two tests up to around 0.4 mm, where-after variation of the two profiles is observed beyond 0.4 mm. Since no abnormalities were experienced during IHD execution, the repeatability within the first 0.4 mm suggests that the tests were executed consistently; therefore stress field variations may actually exist within the sample. The IHD results of test B shown in Fig. 2 are subsequently used to compare to other methods as the SXRD measurements were performed on the same sample. In addition, the hole eccentricity for sample B was lower at 0.02 mm compared to test A of 0.06 mm which implies the results of test A fall outside of the ASTM E837-13 specifications.

The CM results are depicted in Fig. 3. In order to make comparisons to the other techniques as shown in Fig. 4, a line profile down the centre is extracted, and the stresses are averaged over a 2 mm area (as indicated by the arrows in the top image of Fig. 3). The stress averaging provides similar dimensions to the SXRD and IHD measurements. The graph in Fig. 3 provides the

48

extracted stress profiles over the 2 mm centre region which appears as a series of continuous lines. The averaged stress over the ±1 mm is indicated by the black round data points. The variation of the stress field is around ±12 MPa in the first 0.4 mm from the surface, with increasing variations with depth to around ±25 MPa from 0.9 to 1.5 mm. Although there are uncertainties associated with each measurement technique, the variation observed by CM within a single sample supports the notion of stress field non-uniformities suggested upon evaluation of the SXRD results. The averaged profile indicates a depth of compressive stress in the range of 1.1 to 1.2 mm.

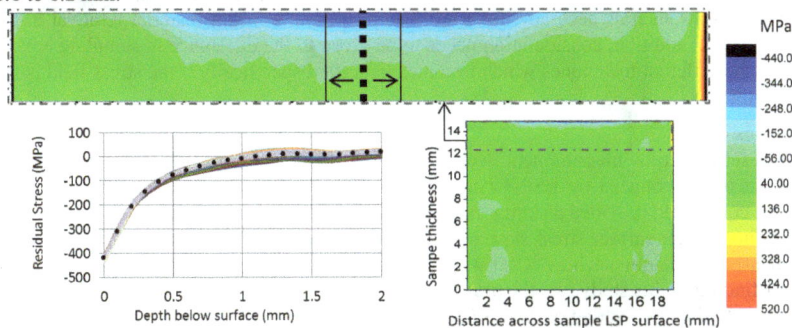

Fig. 3: Residual stress cross-sectional map of the LSP sample using the contour method on the full sample (right), a view of the first 2 mm (top), and a stress profile extracted from the centre of the sample ±1 mm (mean over ±1 mm stress indicated by round black points).

Comparing the XRD and SXRD measurements as depicted in Fig. 4, the surface stress state is similar, however the XRD depth of compressive stress obtained by electro-polishing results in a transition to tension at around 0.8 mm compared to beyond 1 mm for the SXRD results as seen in Fig 2b. Although this is plausible, especially considering the existence of a non-uniform stress field as revealed by the contour measurement, it is possible that the material removal correction factor needs to be evaluated, as only the central section of the coupon has been LSP processed although the entire surface is removed by electro-polishing.

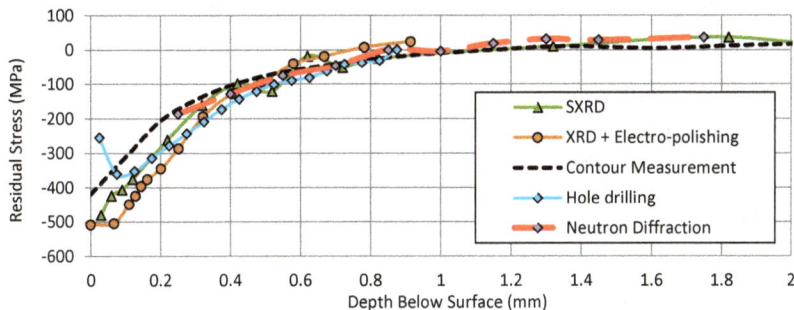

Fig. 4: A comparison residual stress profiles measured on the LSP sample using the different measurement techniques covered in this manuscript.

MECA SENS 2017   Materials Research Forum LLC
Materials Research Proceedings **4** (2018) 45-50   doi: http://dx.doi.org/10.21741/9781945291678-7

The mean stress profile of the contour measurement and the ND results both agree well with the SXRD with respect to the indicated depth of compressive residual stress. The surface residual stress state indicated by XRD is -508 MPa, however the surface stress state found by the CM method is slightly lower at -419 MPa. The XRD results shown in Fig. 2a revealed a range of around 100 MPa (-480 to -580 MPa), therefore the surface stress result found using the contour method is not a significant variation. However, both IHD tests revealed a significantly underestimated surface stress state (of around -255 MPa) compared to the other techniques. Typically the IHD technique is not as well suited to surface stress state quantification for a number of reasons such as accuracy in surface detection, and uniform material removal in the first few steps. In addition, consideration needs to be given to the difference in sampling area or volume of the different techniques which becomes important specifically if the stress field is not perfectly uniform.

Conclusions

Various residual stress measurement techniques have been employed in order to quantify residual stresses introduced by LSP for applications of LP steam turbine blades. Each of the techniques considered provided a similar indicated depth of compressive residual stress, however variations in the near surface stress state were observed which highlights the value of utilizing a complimentary technique such as XRD which is a better suited surface stress measurement. A particular finding of this study was the presence of a potentially non-uniform stress field which is possibly due to the low overlap of the LSP spot array. Variations in the stress field were detected using fine resolution offered by SXRD and a full field map by the contour measurement. It is therefore beneficial to consider the potential stress averaging effect associated with a particular measurement technique as a selected technique may not necessarily reveal stress field non-uniformities.

Acknowledgements

The authors would like to acknowledge key contributions of Dr Thomas Buslaps at the ESRF, and Jeferson Oliveira of the Open University regarding contour measurements. In addition, the authors are grateful to Dr Axel Steuwer for GSAS data analysis regarding SXRD measurements.

References

[1] K. Ding and L. Ye, Laser shock peening performance and simulation, Woodhead Publishing Limited, Cambridge, England, 2006. https://doi.org/10.1201/9781439823620

[2] M.N. James, M. Newby, D.G. Hattingh and A. Steuwer, Shot-peening of steam turbine blades: residual stresses and their modification by fatigue cycling, Procedia Eng. 2.1 (2010) 441-451. https://doi.org/10.1016/j.proeng.2010.03.048

[3] M.G. Moore and W. P. Evans, Mathematical correction for stress in removed layers in X-ray diffraction residual stress analysis. No. 580035. SAE Technical Paper, 1958.

[4] P.V. Grant, J.D. Lord and P.S. Whitehead, The Measurement of Residual Stresses by the Incremental Hole Drilling Technique. NPL Materials Centre. Measurement Good Practice. Guide. 53 (2006)

[5] M. Newby, D. Glaser and C. Polese, Laser Shock Peening Process Development for Turbine Blade Refurbishment Applications Using a Commercial "Mid-Range" Energy Laser", 6th ICLPRP, Skukuza, SA. (2006)

[6] M. T. Hutchings, P. J. Withers, T. M. Holden and T. Lorentzen, Introduction to the characterization of residual stress by neutron diffraction, CRC press, 2005.

Microstructure Characterisation

MECA SENS 2017
Materials Research Proceedings 4 (2018) 53-58

Materials Research Forum LLC
doi: http://dx.doi.org/10.21741/9781945291678-8

Diffraction Line-Broadening Analysis of Al$_2$O$_3$/Y-TZP Ceramic Composites by Neutron Diffraction Measurement

K. Fan[1,a], J. Ruiz-Hervias[2,b*], C. Baudin[3,c] and J. Gurauskis[3]

[1]College of Materials Science and Engineering, Xihua University, Chengdu 610039, China

[2]Materials Science Department, Universidad Politécnica de Madrid, E.T.S.I. Caminos, Canales y Puertos, Profesor Aranguren 3, E-28040 Madrid, Spain

[3]Instituto de Cerámica y Vidrio, CSIC, Kelsen 5, E-28049 Madrid, Spain

[a]fankunyang123@126.com, [b]jesus.ruiz@upm.es, [c]cbaudin@icv.csic.es

Keywords: Ceramic Composites, Tape Casting, Neutron Diffraction, Peak Broadening, Microstrain

Abstract. Time-of-flight neutron diffraction was used for through-thickness measurement in a series of Al$_2$O$_3$/Y-TZP (alumina/tetragonal ZrO$_2$ stabilized with 3 mol.% Y$_2$O$_3$) ceramic composites. Different zirconia contents (5 vol.% and 40 vol.%) and green processing routes (a novel tape casting and conventional slip casting) were investigated. Diffraction line broadening analysis was carried out by using the Rietveld refinement method combined with the "double-Voigt" modelling to obtain the domain size and microstrain in the different phases. The results indicate that peak broadening is noticeable in zirconia (associated to microstrains) but not in alumina. The microstructure and non-uniform microstrains were mainly influenced by the Y-TZP content in the studied Al$_2$O$_3$/Y-TZP composites, irrespective of the measured direction and the fabrication process.

Introduction

Alumina-zirconia ceramics have received considerable attention in both engineering and academic fields due to their improved mechanical properties compared with pure alumina ceramics. Most of the previous research works [1] were focused on the uniform residual stresses between phases, without paying too much attention to the non-uniform microstrains at the grain and subgrain scales (type III stresses), maybe due to the tougher requirements both on the measurement and analysis procedures. Such non-uniform microstrains can reflect the existence of crystal defects (e.g., dislocations and crystal vacancies), which in turn will influence the mechanical properties of materials. Consequently, it is important to quantify the non-uniform microstrains of ceramics by using suitable measurement techniques and analysis methods.

Neutron diffraction is a preferred measurement technique to get information inside bulk samples, due to its high penetration depth. Based on the diffraction measurements, microstructural information of materials, e.g., crystal size and microstrain, can be extracted by diffraction line broadening analysis. Several approaches were established for quantification of line-broadening effects, e.g., the simplified integral-breadth methods, the Warren-Averbach procedure [2], and the traditional Williamson-Hall method [3]. Each method has its limitations and advantages, and sometimes conflicting results are obtained by different methods. Compared with other methods, the 'double-Voigt' approach [4] combined with Rietveld refinement shows its advantages in line broadening analysis for neutron data, especially in cases with limited *a priori* information, e.g., an arbitrary sample where significant peak overlap occurs.

In the present work, a series of Al$_2$O$_3$/Y-TZP (alumina/tetragonal ZrO$_2$ stabilized with 3 mol.% Y$_2$O$_3$) ceramic composites were studied. A novel tape casting route [5], which is better than

MECA SENS 2017   Materials Research Forum LLC
Materials Research Proceedings **4** (2018) 53-58   doi: http://dx.doi.org/10.21741/9781945291678-8

traditional casting methods both from the economic and environmental point of view, was used to fabricate the ceramic composites samples. Samples obtained with the conventional slip casting technique [6] were taken as reference for comparison. The time-of-flight (TOF) neutron diffraction technique was used for measurement in the studied ceramic composites. Line broadening analysis was carried out to obtain the microstructural information (domain size and crystal microstrain).

Materials and Methods

Materials. The studied Al_2O_3/Y-TZP (3 mol.% Y_2O_3 stabilized zirconia) ceramic composites were prepared by two different green processing methods: the novel tape casting [5], and the conventional slip casting, as a reference technique. Two different contents of Y-TZP were used as reinforcement: 5 vol.% and 40 vol.%. The studied specimens were coded as A-5YTZP (slip), A-5YTZP (tape), A-40YTZP (slip), and A-40YTZP (tape), to describe their composition and fabrication technique.

High-density (relative density > 98 % theoretical density) materials were obtained after sintering for all the studied composites. The morphology of the chemical etched surface and the fracture surface of samples were observed by scanning electron microscopy (SEM, Zeiss DSM 950, Germany) and the average grain sizes of alumina and zirconia particles were determined.

Neutron Diffraction Measurements. For neutron diffraction strain scanning, sintered samples in the shape of parallelepipeds with dimensions $20 \times 20 \times 5$ mm^3 were employed. Neutron diffraction data were collected on ENGIN-X time-of-flight instrument [7], at the ISIS, UK. The experimental setup consists of two detector banks which are centered on Bragg angles of $2\theta_B = \pm 90$ degrees. This setup allows simultaneous measurements in two directions of the sample: the in-plane (contained in the stacking plane) and the normal one (along the sample thickness). Through-thickness scanning was carried out along the sample thickness (approximately 5 mm), in 0.4 mm steps.

Data Analysis. The whole diffraction pattern was obtained in TOF diffraction measurements and then analyzed by Rietveld refinement, using the TOPAS-Academic V5 software package [8]. The diffraction profile is a convolution of instrumental and sample effects, where the latter is a combination of size and strain broadening. The overall quality of fitting was assessed in terms of R values [9], as obtained from the refinements.

In order to accurately determine the physical broadening, a correction of instrument broadening is essential. CeO$_2$ standard powder (cubic phase, space group $Fm\bar{3}m$, $a = 5.4114$ Å), as a highly crystalline sample which brings minimum sample broadening, was used as standard material for instrumental calibration. Good fits were achieved for the patterns of the CeO$_2$ standard powder recorded in each detector bank, with the weighed residual error R_{wp} ranging from 4 % to 8 %. The instrument-dependent parameters obtained from the fitting were fixed for subsequent refinement of the profiles corresponding to the ceramic composite samples.

After instrumental calibration, the sample effects, i.e., microstrain (strain broadening) and small crystallite size (size broadening), were analyzed by "double-Voigt" modelling. The crystallite size and microstrain values were determined [4], i.e., as volume-weighted domain size D_V and the average microstrain e, respectively. Details of the refinement procedure are given in [10].

Results and Discussion

Microstructure. Very similar microstructures were found for the studied Al_2O_3/Y-TZP samples with the same composition for both fabrication techniques, i.e. slip casting and tape casting. Fig. 1 shows the SEM micrograph of chemical etched surface and the fractured surface of the tape-cast Al_2O_3/Y-TZP composites, as a function of zirconia content. X-ray diffraction analysis previously reported [11] that only α-Al_2O_3 and tetragonal Y_2O_3-stabilized zirconia (Y-TZP) phases were

MECA SENS 2017   Materials Research Forum LLC
Materials Research Proceedings **4** (2018) 53-58   doi: http://dx.doi.org/10.21741/9781945291678-8

found in the sintered ceramics. It can be seen that the alumina matrix (in dark grey) and zirconia particulates (in light grey) were generally well-dispersed in all the studied materials. Due to the inhibition effect of the second phase zirconia particles on matrix grain growth, the grain size of the Al_2O_3 matrix was decreased as the zirconia content increased from 5 vol.% to 40 vol.%, with the average value dropping from 1.9 ±0.3 μm to 1.1 ±0.2 μm (compare Fig. 1a with 1b). Narrow grain size distributions were observed for both phases in each composite.

Fig. 1: SEM micrographs of chemical etched surfaces and fractured surface (the inset) of the studied tape casting materials: (a) A-5YTZP (tape) composites. (b) A-40YTZP (tape) composites. Al_2O_3 grains appear with dark grey color and Y-TZP particulates are in light grey color.

Peak broadening analysis. Diffraction profiles of samples were fitted without (only instrument broadening) and with sample physical broadening effect, respectively. As an example, the profile fitting of the A-40YTZP (tape) sample is presented in Fig. 2. The observed (measured) peak profile is shown as a blue line. The difference plot (difference between the observed and calculated intensities), is shown in grey below the spectra. The individual peaks of α-Al_2O_3 and Y-TZP phases were identified with blue and black tick marks, respectively, below the profile. The strongest peaks of the Al_2O_3 (113) and Y-TZP (112) & (200) phases are shown in detail, in Figs. 2b and 2c, respectively.

If only the instrument effect is considered, as shown by the green line in Fig. 2, the sample contribution is not captured. The overall quality of fitting of the full profile with only instrument effect is far from optimum, with the weighted residual error R_{wp} ranging from 14~17 %, and the corresponding difference curves (in grey color below the spectra) showing remarkable fluctuations. The calculated peak profiles of Y-TZP reflections are narrower than the measured ones (Fig. 2c). However, such behavior is not noticeable in the Al_2O_3 reflections. With only instrument effect, the calculated Al_2O_3 peak profiles fitted the measured one quite well, as shown in Fig. 2b.

After introducing the sample physical contribution, i.e., crystallite size and microstrain, much better fitting (red line) was achieved, as can be judged from the flatter difference curves (black line below the spectra), as well as the lower R_{wp} = 6 %. All of the Y-TZP and Al_2O_3 reflections were well fitted when the physical contribution to line broadening was taken into account.

Such peak broadening was observed in the Y-TZP reflections in all investigated samples, but it is almost negligible for the Al_2O_3 matrix. No obvious anisotropic broadening was observed in the profile fitting, which indicates the absence of texture or preferred orientation in the samples.

The "double-Voigt" modelling procedure for Al_2O_3 reflections gave unrealistic values of the size- and strain-related parameters, with very large error values. This indicates that there is no size or strain broadening in the Al_2O_3 phase, in agreement with the profile fitting behavior presented above. As a stable phase, negligible microstrains were reported for alumina in [12]. On the other hand, above a certain crystallite size (~1 μm), size broadening is almost negligible in diffraction techniques. For the studied composites, the average grain size of the Al_2O_3 matrix was generally larger than one micrometer (Fig. 1), and peak broadening was hardly observed.

For the Y-TZP phase, according to the analysis, peak broadening was mainly due to strain while

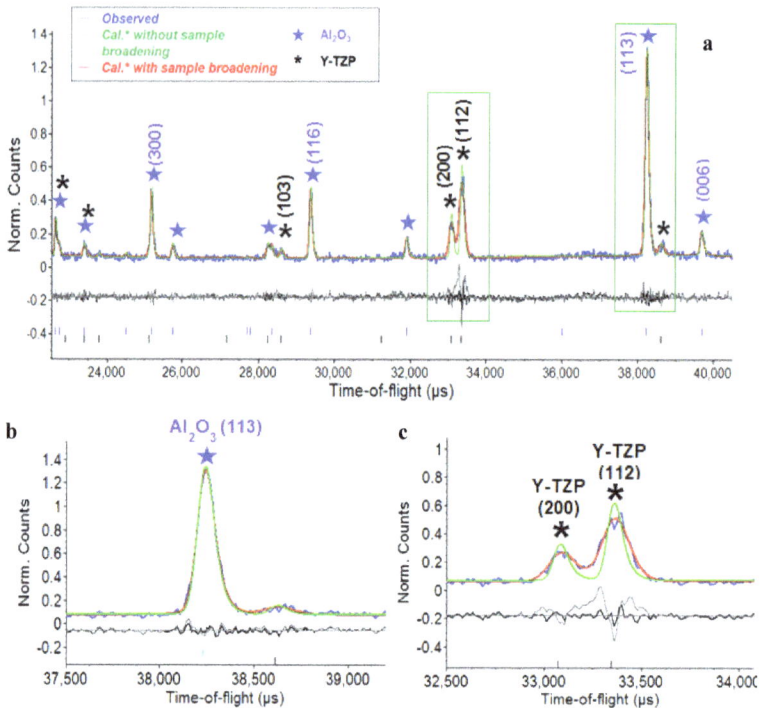

Fig. 2: Profile fitting for A-40YTZP (tape) sample: observed peak profile, blue line; fitting without physical broadening, green line, R_{wp} = 15 %; fitting with broadening, red line, R_{wp} = 6 %. (a) full profile: (b) Al_2O_3 (113) reflection; (c) Y-TZP (112) & (200) reflections.

size broadening effects were negligible. This might be explained by considering the zirconia phase transformation during fabrication. For the studied samples, 30 vol.% monoclinic ZrO_2 was found in the initial powders that was completely transformed to tetragonal phase during sintering [13]. It was reported [14] that atomic displacements occurred during the monoclinic to tetragonal zirconia phase transition in the alumina-zirconia system, mainly involving oxygen atoms. Some authors claim that a shearing mechanism also happens in the zirconia phase transformation [15]. In addition to that, an inhomogeneous distribution of yttria was observed during sintering in another work [12] with similar materials, which induces a defective core-shell structure of Y-TZP grains. All these factors could give rise to the observed microstrain in the Y-TZP.

The average microstrain (e) values of Y-TZP are presented in Fig. 3. The data correspond to average values over different scanned positions in each sample, with standard errors represented by error bars. As the Y-TZP vol.% increases, a slight increase in microstrain e in the Y-TZP crystallite was detected, changing from around $4 \cdot 10^{-4}$ in the A-5YTZP to $6 \cdot 10^{-4}$ in the A-40YTZP composites. This agrees with the findings of Wang *et al.* [16] and A. Reyes-Rojas *et al.* [17], which previously reported a linear increase in microstrain with an increase in ZrO_2 % in the Al_2O_3-ZrO_2 composites. The increase in microstrain in the Y-TZP crystallite would indicate that the number of lattice defects increases with the Y-TZP content. In addition, the obtained microstrain might provide an additional increment of local strain which would result in crack initiation and propagation.

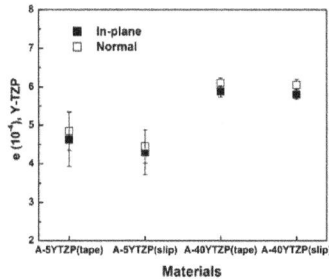

Fig. 3: The average microstrain, e, in Y-TZP in all of the studied Al₂O₃/Y-TZP composites.

No significant difference was detected in the calculated microstrain of the Y-TZP due to the different manufacturing techniques employed in this work (tape casting and slip casting). The slight differences between the A-5YTZP (slip) and the A-5YTZP (tape) are included in the error bars, as shown in Fig. 3. Differences between the in-plane and the normal directions were also included within the experiment error bars for all composites samples.

Summary

A series of Al_2O_3/Y-TZP bulk composites fabricated by different green processing (the novel tape casting and conventional slip casting) and with different Y-TZP content were investigated by time-of-flight neutron diffraction. Diffraction line broadening analysis was carried out by using the Rietveld refinement method to extract the microstructural information in the sample. Peak broadening in the Y-TZP reflections was observed in all investigated A/Y-TZP composites, but not in the Al_2O_3 reflections. The line-broadening of the Y-TZP peaks was mainly due to non-uniform microstrains. By increasing the Y-TZP content in the A/Y-TZP composites, the non-uniform microstrain e in the Y-TZP crystallite increased from around $4 \cdot 10^{-4}$ in the A-5YTZP to $6 \cdot 10^{-4}$ in the A-40YTZP composites. No obvious difference in microstructure and peak broadening was observed due to sample orientation (in-plane and normal directions) and the manufacturing processes (the novel tape casting and the conventional slip casting). Consequently, the quality of the ceramics manufactured using the novel tape casting method was proven (*e.g.*, homogeneous microstructure, without undesired defects induced by the green processing method.). The quality of the ceramics manufactured using the novel tape casting method was proven (e.g., homogeneous microstructure, without undesired defects induced by the green processing method).

References

[1] M.E. Fitzpatrick, A.T. Fry, P. Holdway, F.A. Kandil, J. Shackleton and L. Suominen, Measurement Good Practice Guide No. 52 - Determination of Residual Stresses by X-ray Diffraction - Issue 2 (2005)

[2] B. Warren and B. Averbach, The separation of cold work distortion and particle size broadening in X-ray patterns, J. Appl. Phys. 23 (1952) 497-497. https://doi.org/10.1063/1.1702234

[3] G. Williamson and W. Hall, X-ray line broadening from filed aluminium and wolfram, Acta Metall. 1 (1953) 22-31. https://doi.org/10.1016/0001-6160(53)90006-6

[4] D. Balzar, N. Audebrand, M. Daymond, A. Fitch, A. Hewat, J. Langford, A. Le Bail, D. Louër, O. Masson and C. McCowan, Size-strain line-broadening analysis of the ceria round-robin sample, J. Appl. Crystallogr. 37 (2004) 911-924. https://doi.org/10.1107/S0021889804022551

[5] J. Gurauskis, A. Sanchez-Herencia and C. Baudin, Joining green ceramic tapes made from water-based slurries by applying low pressures at ambient temperature, J. Eur. Ceram. Soc. 25 (2005) 3403-3411. https://doi.org/10.1016/j.jeurceramsoc.2004.09.008

[6] A. Tsetsekou, C. Agrafiotis and A. Milias, Optimization of the rheological properties of alumina slurries for ceramic processing applications Part I: Slip-casting, J. Eur. Ceram. Soc. 21 (2001) 363-373. https://doi.org/10.1016/S0955-2219(00)00185-0

[7] J. Santisteban, M. Daymond, J. James and L. Edwards, ENGIN-X: a third-generation neutron strain scanner, J. Appl. Crystallogr. 39 (2006) 812-825. https://doi.org/10.1107/S0021889806042245

[8] A. Coelho, TOPAS-Academic V5, Coelho Software, Brisbane, Australia, http://www.topas-academic.net/ (2012)

[9] L. McCusker, R. Von Dreele, D. Cox, D. Louer and P. Scardi, Rietveld refinement guidelines, J. Appl. Crystallogr. 32 (1999) 36-50. https://doi.org/10.1107/S0021889898009856

[10] K. Fan, J. Ruiz-Hervias, J.Y. Pastor, J. Gurauskis and C. Baudín, Residual stress and diffraction line-broadening analysis of Al2O3/Y-TZP ceramic composites by neutron diffraction measurement, Int. J. Refract. Metals Hard Mater. 64 (2017) 122-134. https://doi.org/10.1016/j.ijrmhm.2017.01.011

[11] J. Gurauskis, Desarrollo de materiales laminados de alúmina-circona reforzados por tensiones residuales, (2006). Ph.D. dissertation (in Spanish)

[12] C. Exare, J.M. Kiat, N. Guiblin, F. Porcher and V. Petricek, Structural evolution of ZTA composites during synthesis and processing, J. Eur. Ceram. Soc. 35 (2015) 1273-1283. https://doi.org/10.1016/j.jeurceramsoc.2014.10.031

[13] K. Fan, J. Ruiz-Hervias, J. Gurauskis, A.J. Sanchez-Herencia and C. Baudín, Neutron diffraction residual stress analysis of Al2O3/Y-TZP ceramic composites, Boletín de la Sociedad Española de Cerámica y Vidrio, 55 (2016) 13-23. https://doi.org/10.1016/j.bsecv.2015.10.006

[14] D. Simeone, G. Baldinozzi, D. Gosset, M. Dutheil, A. Bulou and T. Hansen, Monoclinic to tetragonal semireconstructive phase transition of zirconia, Phys. Rev. B 67 (2003). https://doi.org/10.1103/PhysRevB.67.064111

[15] R. Patil and E. Subbarao, Monoclinic-tetragonal phase transition in zirconia: mechanism, pretransformation and coexistence, Acta Crystallogr. A 26 (1970) 535-542. https://doi.org/10.1107/S0567739470001389

[16] X.L. Wang, C.R. Hubbard, K.B. Alexander, P.F. Becher, J.A. Fernandez-Baca and S. Spooner, Neutron diffraction measurements of the residual stresses in Al2O3-ZrO2 (CeO2) ceramic composites, J. Am. Ceram. Soc. 77 (1994) 1569-1575. https://doi.org/10.1111/j.1151-2916.1994.tb09758.x

[17] A. Reyes-Rojas, H. Esparza-Ponce, S.D. De la Torre and E. Torres-Moye, Compressive strain-dependent bending strength property of Al2O3-ZrO2 (1.5 mol% Y2O3) composites performance by HIP, Mater. Chem. Phys. 114 (2009) 756-762. https://doi.org/10.1016/j.matchemphys.2008.10.044

MECA SENS 2017
Materials Research Proceedings **4** (2018) 59-64

Materials Research Forum LLC
doi: http://dx.doi.org/10.21741/9781945291678-9

Influence of Laser Power and Traverse Speed on Weld Characteristics of Laser Beam Welded Ti-6Al-4V Sheet

P.M. Mashinini[1,a*] and D.G. Hattingh[2,b]

[1]University of Johannesburg, Department of Mechanical and Industrial Engineering Technology, Doornfontein Campus, Johannesburg, South Africa

[2]Nelson Mandela University, Department of Mechanical Engineering, North Campus, Summerstrand, Port Elizabeth, South Africa

[a]mmashinini@uj.ac.za, [b]danie.hattingh@mandela.ac.za

Keywords: Laser Beam Welding, Ti-6Al-4V, Traverse Speed, Microstructure, Residual Stress

Abstract. In this paper laser beam welding was used for joining 3 mm Ti-6Al-4V alloy sheets in a full penetration butt-weld configuration. Laser beam power and traverse speed were the only parameters varied in an attempt to characterize the influence on weld integrity with specific reference to residual stress and microstructural modifications. The iXRD residual stress data showed a definite influence of traverse speed on residual stresses, with low traverse speeds resulting in an increased tensile residual stresses in the longitudinal direction of the weld whilst in the transverse direction residual stress revealed a more compressive stress state. The residual stress data for this experiment compared favourably with published residual stress data done by synchrotron X-ray diffraction. Weld joint integrity was further analyzed by evaluating the microstructure transformation in the weld nugget. These results revealed a degree of grain growth and the presences of fine acicular β (needle-like α) in prior β grain boundaries with increased traverse speed. Grain growth was predominantly influenced by the cooling rate which is associated with traverse speed. Additionally, the α -phase and β -phase were characterized in the various weld zones by electron backscatter diffraction (EBSD).

Introduction

Laser Beam Welding (LBW) as a fusion joining technique which was invented in the late 1960s [1], has successfully been utilized to weld light metal alloys including aluminium, magnesium and titanium [1, 2]. LBW can be used to weld linear or rotational joints using a carbon dioxide (CO_2) laser. Alternatively, complex joints can be made using a neodymium-doped yttrium aluminium garnet (Nd:YAG) laser [2, 3, 4]. During LBW, a high power laser is focused on the material weld joint-line (typically 1 kW to 5 kW) [2, 5]. LBW process involves heat conduction and heat absorption on the workpiece on the joint-line. This causes material to melt. The temperature increases above the boiling point of the parent material. The vapour pressure increases, creating a narrow capillary (keyhole) which propagates through the material. The keyhole traps almost all the laser power. The keyhole is then filled with metallic vapour as the laser beam traverses, forming a welded joint [6, 7].

Literature has shown some research has been done on LBW processing and it is a growing field. Liu *et al.* [8] reported research work on fatigue damage evolution on pulsed Nd:YAG Ti-6Al-4V laser welded joints. The results indicated a martensitic (α') phase and an underfill flaw in the weld nugget [8]. The results are similar those reported by Xu *et al.* [9] on microstructure characterization of laser welded Ti-6Al-4V fusion zones. The results predominantly showed the martensitic (α') phase with α and retained β phase. The cooling rate was the main source of phase transformation in the weld nugget, that is, high traverse speed gives high cooling rate [9, 10].

Currently, there are concerns regarding residual stresses (distortion), microstructural transformation and the formation of cavities/porosity in the weld joint due to the exposure to laser power. These factors are a concern for the mechanical performance of the welded components.

Experimental Technique
Laser beam welding was utilised for joining of 3 mm mill annealed Ti-6Al-4V alloy sheet in a full penetration butt-weld configuration. The sheets chemical composition were: (wt.%) Al 6.25, V 4.04, Fe 0.19, C 0.018, N 0.008, O 0.18 and balance Ti. The welding platform used for this research was a TRUMPF LASERCELL 1005 (TLF laser) based at the National Laser Centre (NLC) in Pretoria, South Africa. Weld coupons for LBW were processed with laser power ranging from 2.3 kW to 4.3 kW and varying traverse speed between 1 m/min and 5 m/min. The weld pool was shielded with Argon gas to reduce oxidation of the weld nugget. The welded plates were cut and sectioned for macrostructure and microstructural evaluation. The samples were mounted and etched using a solution of 2 ml HF (40 %), 5 ml H_2O_2 (30 %) and 10 ml H_2O for approximately 30 seconds. An optical microscope and HR-TEM were used to evaluate the microstructure of the welds. Surface residual stresses were measured using an iXRD Residual Stress Measurement System manufactured by Proto Manufacturing Ltd [11]. The measurements were done on the weld sheets in as as-welded condition. Surface residual stresses were measured along and across the weld, which will represent transverse (sT) and longitudinal stress (sL) respectively. A 3 mm circular aperture was used as it was found that it's the only size that gave a satisfactory resolution and higher peak intensity for this research material. This was attributed to the very fine microstructure of Ti-6Al-4V alloy. A spacing of 1 mm was used for measurements from one point to the next, this allowed for good averaging of data points as an aperture size of 3 mm allowed for an overlap between points. A Bragg angle of 139.69° and 5 Beta oscillations per measurement were used. Copper tube and Nickel filters were used for this research as they are the recommended materials to use when measuring residual stresses for Titanium alloys [11]. For all measurements, an elliptical curve fitting was used for the residual stress data which allowed for the determination of the normal and shear stress components. But only the normal stress component is reported in this research as the shear stress component was found to be very small. The parent plate yield and tensile strength is 890 MPa and 1017 MPa respectively.

Results
There was a flaw-like indications observed in these welds made as show in Table 1. The weld made at welding speed of 1 m/min showed evidence of severe undercut at the weld root. A possible explanation for this is the sudden contraction of the near molten material during cooling of the weld pool. In general, most welds had indications of undercut at the bottom of the weld.

Table 1: Macrographs of welds.

Laser Power [kW]	2.3	3.3			4.3
Welding Speed [m/min]	3	1	3	5	3
Weld cross-section					

MECA SENS 2017   Materials Research Forum LLC
Materials Research Proceedings **4** (2018) 59-64   doi: http://dx.doi.org/10.21741/9781945291678-9

Microstructure

From the optical micrographs, the parent plate showed an α phase in a matrix of retained β phase. The HAZ showed transformed β phase containing acicular α phase. But lower traverse speed resulted in grain growth as compared to high traverse speed. In the weld nugget, low traverse speed resulted in coarse acicular α phase (needle-like α) in prior β grain boundaries while high traverse speed resulted in fine acicular α phase (needle-like α) in prior β grain boundaries. The optical micrographs for the weld zones are illustrated in Table 2. Additionally, the HAZ and weld nugget size varied from wide to narrow as the traverse speed is increased. This was mainly attributed to the cooling rate after the welding process.

Table 2: Indicative optical micrographs for the various weld zones.

Traverse Speed [m/min]	Weld Nugget	Transition Zone	Heat Affected Zone
1			
3			
5			

The microstructure was further characterised by placing the samples in the Scanning Electron Microscope (SEM) by first assessing the parent plate to identify the α/β phase distribution of the Ti-6Al-4V alloy. By utilising electron backscatter diffraction (EBSD), the parent plate contained approximately 90 percent α phase (6-10 μm) and about 7 percent retained β phase (1-2 μm). Figure 1 shows the SEM and EBSD images with α phase indicated in blue and β phase in red.

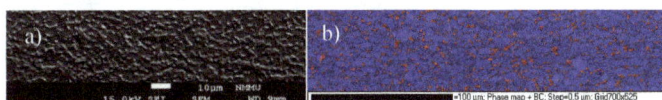

Fig. 1: a) SE SEM images and b) EBSD phase maps (α-blue: β-red) of parent plate.

Further EBSD analysis was done on the welds. The evaluated areas were the weld nugget and the heat affected zones respectively. The weld nugget for all welds showed similar results, there were no traces of β phase irrespective of the traverse speed. Figure 2 to Fig. 3 shows the EBSD phase maps for the welds done. This results are similar to those achieved by Steuwer *et al.* [12] when evaluating the friction stir welded Ti-6Al-4V alloy [12]. Further EBSD analysis was done on the welds. The evaluated areas were the weld nugget and the heat affected zones respectively. The weld nugget for all welds showed similar results, there were no traces of β phase irrespective of the traverse speed. The Scanning Electron Microscope (SEM) was used for crystallographic orientation distribution first for the Ti-6Al-4V alloy parent plate using Electron Backscattered Diffraction (EBSD). The parent plate showed a $\{0001\}<001>$ plane as the preferred/dominating as shown in Table 3. At low traverse speed, there was no clear preferential

MECA SENS 2017   Materials Research Forum LLC
Materials Research Proceedings **4** (2018) 59-64   doi: http://dx.doi.org/10.21741/9781945291678-9

orientation from the HAZ to the weld nugget as indicated in Table 4. Welding direction (WD) is also shown. At intermediate speed, the dominating slip system in the weld nugget and HAZ was between {0001}<001> and {1120}<120> respectively. The weld nugget at high traverse speed showed preferential direction of {1010}<010>. Additionally for both the HAZ and weld nugget, there was evidence of grain growth. The weld nugget at intermediate traverse speed showed prismatic deformation or primary slip at {1120}<120> which has the lowest shear stress compared to basal and pyramidal planes. This slip plane is an indication to have a better performance for fatigue testing.

Table 3: EBSD phase maps (α-blue: β-red) of 3.3 kW at 1 m/min, 3 m/min, 5 m/min for HAZ and Weld Nugget.

Laser Power	Traverse Speed	HAZ	WELD
3.3 [kW]	1 [m/min]		
	3 [m/min]		
	5[m/min]		

Table 4: Inverse pole figure map of 3.3 kW at 1 m/min, 3 m/min, 5 m/min for HAZ and Weld Nugget.

Laser Power	Traverse Speed	HAZ	WELD
Parent Plate			
3.3 [kW]	1 [m/min]		
	3 [m/min]		
	5[m/min]		

Surface Residual Stress

The induced residual stresses data in conjunction with microstructural results was used to assist in explaining the effect of traverse speed on weld integrity. Surface Residual Stress data was obtained by X-ray diffraction technique. The measurements were done 40 mm from the center of the weld on each side. From each point measurement, two readings are recorded, that is, longitudinal and transverse stresses. The principle of measuring surface residual stress is based on Bragg's law, which was established by WL Bragg in 1913 [11]. Diffraction only happens when the material measured has a crystalline structure [11].

Residual stresses using diffraction can be calculated using a number methodologies but for this research, the $\sin^2\psi$ method was used as the mostly used method to calculate stress especially at different psi tilts angles [11]. For any given lattice spacing the stress is calculated using Eq. 1:

$$\sigma_\phi = \frac{E}{(1+v)sin^2\psi}\left(\frac{d_\psi-d_o}{d_o}\right). \tag{1}$$

Where; σ_ϕ= Single stress acting in a chosen direction i.e. ϕ (MPa), E = Elastic modulus (GPa), v = Poisson's ratio, ψ = Angle between the normal plane of specimen and normal of the diffracted plane (°), d_o = Inter-planar spacing at free strain (Å) and d_ψ = Inter-planar spacing of planes at angle ψ to the surface (Å). The parent plate was first measured for residual stress. The results revealed that the plates received from the manufacture had the average compressive stresses of 260 ±21 MPa and 160 ±24 MPa in transverse and longitudinal to the rolling directions respectively. The plates were used as received; no normalising of stresses was done for this research. With respect to the weld, longitudinal direction is along the welding direction and transverse direction is perpendicular to the weld direction. The residual stress results showed high tensile stress in the longitudinal direction for all traverse speeds. The highest stresses recorded was 607 ±62 MPa which was at low traverse speed and low stresses of 422 ±29 MPa at high traverse speeds as indicated in Fig. 2a.

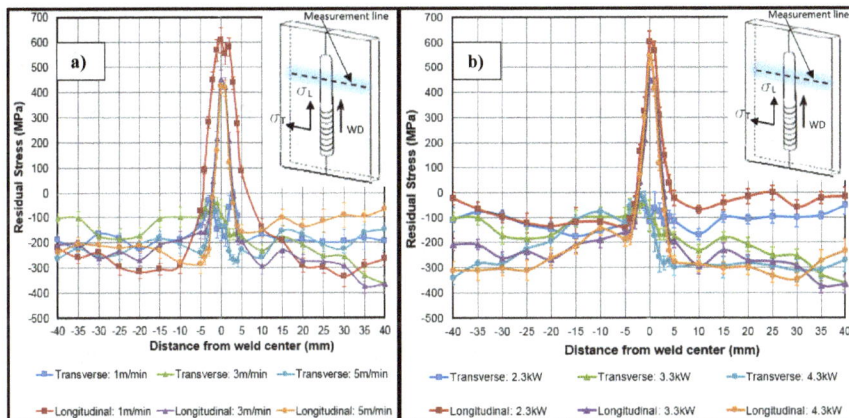

Fig. 2: a) Surface residual stresses for 3.3 kW at 1 m/min; 3 m/min and 5 m/min; b) Surface residual stresses for 3 m/min at 2.3 kW; 3.3 kW and 4.3 kW.

This is as expected at low traverse speed, as the weld nugget becomes wider due to increased time at temperature. There is therefore a larger zone of near molten material in the weld pool. This means that there will be a greater change in width across the weld nugget of a low traverse speed weld versus a high traverse speed weld due to thermal contraction during cooling. This explains why higher residual stresses are found in the weld nugget of low traverse speed welds, even though the low traverse speed welds typically have low cooling rates compared to high cooling rates at high traverse speed. The influence of cooling rate is therefore overwhelmed by the greater influence of thermal contraction of low speeds welds. It therefore follows that as the weld pool cools, a weld with a larger molten pool will experience greater thermal contraction, and hence high residual stresses. Additionally, the transverse direction showed compressive stresses in the range similar to that recorded for the parent plate. Comparing this stresses results with microstructure, it is indicative that the fine microstructure in the weld nugget results in low residual stresses. At traverse speed of 3 m/min, there was no major increase in surface residual stresses with increased laser power. The highest stress was recorded at 600 ±43 MPa at low laser power of 2.3 kW and lowest stress was 545 ±36 MPa at high laser power of 4.3 kW as illustrated in Fig. 2b.

Conclusion

This research showed that traverse speed has an influence in microstructural transformation. Increased traverse speed resulted in grain growth in the weld nugget and HAZ. The nugget also resulted in fine microstructure. High tensile residual stresses were recorded at low traverse speed but no clear indication of increase /decrease in stresses was recorded with increased laser power. Although there was grain growth and fine microstructure in weld nugget with increased traverse speed, this resulted in reduced residual stresses.

Acknowledgements

The authors wish to give thanks to the staff members from Nelson Mandela University and the NLC situated at the CSIR. The National Research Foundation (NRF) for funding provided.

References

[1] A. O'Brien and C. Guzman, Welding handbook: Welding processes, Part 2, 9th ed., American Welding Society. Miami, 2007.

[2] J.C. Ion, Laser processing of engineering materials: Principles, procedure and industrial application, Elsevier, 2005.

[3] C.T. Dawes, Laser welding: A practical guide, Woodhead Publishing, (1 October 1992).

[4] R. Braun,C. Dalle Donne and G. Staniek, Laser beam welding and friction stir welding of 6013-T6 aluminium alloy sheet. Materialwissenschaft und Werkstofftechnik. 31 (12) (2000) 1017-1026. https://doi.org/10.1002/1521-4052(200012)31:12%3C1017::AID-MAWE1017%3E3.0.CO;2-P

[5] E. Akman, A. Demir, T. Canel and T. Sinmzçlik, Laser welding of Ti6Al4V titanium alloys. J. Mater. Process. Technol. 209 (8) (2009) 3705-3713. https://doi.org/10.1016/j.jmatprotec.2008.08.026

[6] L. Migliore, Welding with lasers. Laser Kinetics. (1998)

[7] L. Reclaru, C. Susz and L. Ardelean, Laser beam welding. Timisoara Medical Journal. 60(1): p. 86-89 (2010)

[8] J. Liu, X.L. Gao, L.J. Zhang and J.X. Zhang, A study of fatigue damage evolution on pulsed Nd:YAG Ti6Al4V laser welded joints. Eng. Fract. Mech. 117 (2014) 84-93. https://doi.org/10.1016/j.engfracmech.2014.01.005

[9] P. Xu, L. Li and S. Zhang, Microstructure characterization of laser welded Ti-6Al-4V fusion zones. Mater. Charact. 87 (2014) 179-185. https://doi.org/10.1016/j.matchar.2013.11.005

[10] H. Liu, K. Nakata and N. Yamamoto, Microstructural characteristics and mechanical properties in laser beam welds of Ti6Al4V alloy. J. Mater. Sci. 47 (2012) 1460-1470. https://doi.org/10.1007/s10853-011-5931-8

[11] PROTO Manufacturing. An introduction x-ray diffraction residual stress measurement. (2011)

[12] A. Steuwer, D.G. Hattingh, M.N. James, U. Singh and T. Buslaps, Residual Stresses, microstructure and tensile properties in Ti-6Al-4V friction stir welds. Sci. Technol. Weld. Joi. 17(7) (2012) 525-533. https://doi.org/10.1179/136217112X13439160184196

MECA SENS 2017
Materials Research Proceedings 4 (2018) 65-70

Materials Research Forum LLC
doi: http://dx.doi.org/10.21741/9781945291678-10

Microstresses in Thermally Stable Diamond Composites made by High Pressure Infiltration Technique

V. Luzin[1,a*], G. Voronin[2,b], M. Avdeev[1,c] and J. Boland[3,d]

[1]Australian Nuclear Science and Technology Organisation, Lucas Heights, NSW, 2232 Australia

[2]Smith MegaDiamond Inc, Schlumberger Co, Provo, UT 84604, USA

[3]CSIRO Earth Science and Resource Engineering, Pullenvale, QLD 4069, Australia

[a]vladimir.luzin@ansto.gov.au, [b]gvoronin@hotmail.com, [c]max.avdeevn@ansto.gov.au, [d]Jim.Boland@csiro.au

Keywords: Residual Stress, Diamond Composite, Microstress

Abstract. Microstresses in the diamond and SiC phases of the TSDCs (thermally stable diamond composites), produced by the high pressure infiltration technique, were measured using the neutron diffractometer, KOWARI, at the OPAL research reactor. Microstresses are developed as a result of the cooling and pressure reduction from the sintering high temperature and high pressure (HTHP) conditions. Their magnitude is determined by the thermo-mechanical properties of the SiC matrix and diamond grit, pressure and temperature conditions as well as the exact TSDC phase composition. The experimental results were interpreted in terms the "matrix-inclusion" composite model that was used to evaluate the composite structural integrity.

Introduction

Diamonds composites are finding increasing use in mining, manufacturing and civil construction industries as cutting elements and drilling bits of various toolings. Although diamond is extremely brittle with low impact strength, the diamond composites have superior hardness and toughness better suited to these applications.

There are two diamond-based composite systems. So-called PCD (polycrystalline diamond composite) has a certain amount of metal binder phase, usually Co. The use of the metal binder restricts the operational temperature range for PCD to less than 800°C. The second system of diamond composites is TSDC, which extends the temperature range up to 1400°C [1]. TSDC is free of any metal binder, but a ceramic binder is used instead, usually SiC.

There are several methods of producing TSDC, but most commonly the diamond–SiC composites are produced by the liquid Si infiltration techniques involving reactive sintering. They share some common features to form SiC matrix/binder for the diamond particles since they are all based on reactive bonding operations, usually in HPHT conditions. During this operation, molten silicon infiltrates the diamond powder, reacts with the diamond powders that results in the formation of the diamond-SiC composite. On the other hand, the HPHT conditions can vary drastically. In some methods the sintering is carried out in the diamond stability region (~1500°C and ~5.5 GPa), while others operate in the graphite stable region (~1500°C and ~2.0 – 3.5 GPa) as in case of the process patented by Ringwood in 1980s [2]. There are some variations of the method that involve no pressure or very low pressure [3, 4]. In some cases pressure can be as high as 8-10 GPa and temperature as high as 1800-2000°C [5]. While the first two cases were analyzed for stress earlier [6], the later production route, high pressure silicon infiltration technique [5], is the focus of the current investigation.

While thermal stability is a very important property for a tool's performance, other properties such as impact resistance, thermal fatigue, fracture toughness and wear resistance are related to

MECA SENS 2017   Materials Research Forum LLC
Materials Research Proceedings **4** (2018) 65-70   doi: http://dx.doi.org/10.21741/9781945291678-10

the stress state of the material. The residual stress originates from the manufacturing process, which is usually HPHT process, while the applied stress comes from severe mechanical/thermal loading these composites are subjected to during operation.

Since the current TSDC manufacturing processes involve such high temperatures and pressures, the residual stresses in such composite materials arise from (i) *thermal mismatch*, due to the difference in thermal expansion coefficient between the diamond and SiC matrix phase during the quenching to room temperate and (ii) *elastic mismatch*, due to the difference in bulk elastic modulus between diamond and SiC matrix phases during the pressure drop from high to ambient.

The infiltration process of liquid silicon into the diamond green body is a complex phenomenon. It involves percolation of liquid Si, dissolution of carbon into Si, reaction and formation of SiC. This process is particle-size sensitive and in certain conditions can lead to closure of pores in the diamond compact due to the sealing effect of the SiC formation. This effect is more pronounced for smaller particle sizes of the diamond green body and is also affected by the pressure conditions of the process when diamond particles experience fracturing with effective decrease in the particle size. All these processing variables, including even sample shape and size, can eventually lead to formation of specific microstructures (mostly micro-defect structures such as micro-porosity, micro-cracking and residual silicon) as well as residual microstresses.

Samples

In the current study the focus is on the high pressure (8-10 GPa) high temperature (1800-2000°C) variation of the sintering technology [5]. Samples of two geometries were produced under the same high temperature (~1900°C) and high pressure (~9 GPa) conditions. The two sample geometries were (i) cylinders with diameter of 3 mm and height of 7 mm and (ii) 6 mm side triangular prisms with height of 4 mm (Fig. 1). For the cylindrical samples, finer diamond powder was used in conjunction with coarser grain size (30-40 µm) to provide better infiltration for the whole sample height. For the triangular prism samples finer diamond grit (20-30 µm) was used (Fig. 2). With two different diamond particle sizes and two shapes, the percolation conditions might differ, forming, as discussed above, somewhat different micro-mechanical systems characterized also by different stresses.

Two representative samples for each of the geometries were studied. Samples were characterized in the "as-manufactured" condition with no service history.

Fig. 1: Samples geometries.

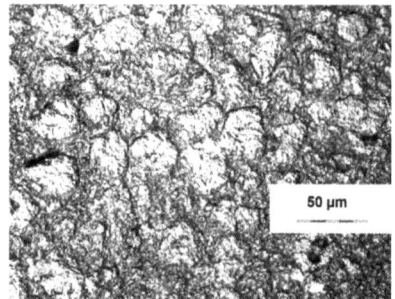

Fig. 2: Microstructure: optical microscopy image of the triangular sample surface.

MECA SENS 2017   Materials Research Forum LLC
Materials Research Proceedings **4** (2018) 65-70   doi: http://dx.doi.org/10.21741/9781945291678-10

Experimental: phase and stress analyses

Two characterizations were conducted by means of neutron diffraction using diffractometers of the Australian research reactor OPAL.

Phase identification and quantitative phase analysis were done at the neutron high-resolution powder diffractometer, WOMBAT [7]. Neutron diffraction patterns were collected in the range of 25° to 145° using wavelength of 1.54 Å and WOMBAT's 120°-range large position sensitive detector. The patterns were analysed using Rietveld refinement technique (GSAS Rietveld refinement software with the EXPGUI interface [8]) to extract volume fractions of diamond and β-SiC (a cubic polymorph, formed at sintering temperatures below 1700°C). No other polymorphs of SiC or C and no residual Si were detected. The accuracy of phase fraction evaluation was better than 0.2 %.

Residual stress analysis was carried out using the neutron residual stress diffractometer, KOWARI [9]. In the residual stress experiment both phases of TSDC samples were measured: diamond (311) reflection and SiC (220) reflection at a wavelength of 1.52 Å. Bragg angles of these reflections were 90° and 58.4° respectively. Due to the overall size of the samples, a relatively small gauge volume with dimension 1.5×1.5×4 mm^3 was used so that measurements could be made from the bulk of the smallest sample. During measurement the gauge volume was positioned in the center of each sample and the sample was rocked in the range of 20° to improve the grain statistics. The measurement time was 10-15 min for the diamond phase and 2-3 hours for the SiC phase.

A special test was made to check isotropy and uniformity of one fine grained (triangular prism) sample by measuring it in multiple directions and several different locations. Since no statistically significant variation in the lattice parameter was found, isotropy and uniformity was confirmed in this way and assumed in the following data analysis.

Determination of the unstressed d-spacing, d_0, is critical for correct microstrain and stress determination in composite materials; therefore it received special attention. The d_0 values for both phases, d_0^{SiC} and d_0^{Dia}, were assumed to be the same in all four samples considering the same treatment history. In principle, it is sufficient to use two equations, corresponding to two samples, setting macrostress to zero, $\sigma_i^{Macro} = f_i^{SiC}\sigma_i^{SiC} + f_i^{Dia}\sigma_i^{Dia} = 0$, $i = 1,2$ in order to resolve the system of two equations with two unknowns, d_0^{SiC} and d_0^{Dia}, and determine them in a unique way (assuming that $f_1 \neq f_2$ to avoid degeneracy). In the above equation f_i^α stands for the volume fraction of the α-th phase and σ_i^α is the total hydrostatic phase stress,

$$\sigma_i^\alpha = \frac{1}{\frac{1}{2}S_2^\alpha(hkl)+3S_1^\alpha(hkl)}\frac{d_i^\alpha-d_0^\alpha}{d_0^\alpha}, \alpha = SiC, Dia.$$

In the case of four samples, the system of such equations is overdetermined, therefore, d_0 values were defined in such a way that provided the condition of zero macrostress in all four samples simultaneously in the least-squares minimization sense (minimizing the

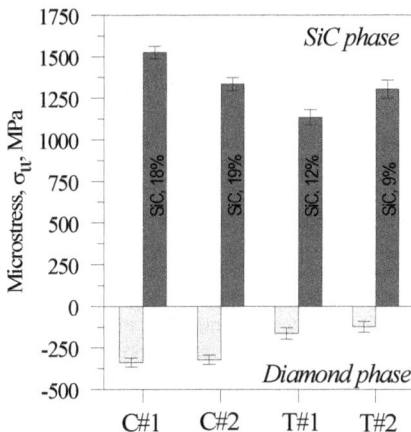

Fig. 3: *Microstress of the SiC and diamond phases in the cylindrical (C) and triangular (T) samples.*

sum of all four squared macrostress residuals). Treated this way, the d_0 for diamond and SiC were determined with accuracy of the same order of magnitude as the accuracy of the measurements, $\sim 1 \times 10^{-5}$ for diamond and $\sim 5 \times 10^{-5}$ for SiC.

Strain-to-stress calculations were carried out using appropriate hkl-dependent diffraction elastic constants calculated from the single crystal elastic constants of diamond and SiC [10]. Numerically, $S_1(hkl)$ and $\frac{1}{2}S_2(hkl)$ were -0.28 and 1.42 TPa^{-1} for the diamond (311) reflection, while for the SiC (220) reflection they were -0.40 and 2.71 TPa^{-1}.

Results and discussion

The results of the neutron phase analysis and neutron stress analysis, based on experimentally determined phase composition are shown in Table 1 and visually illustrated in Fig. 3. With the accuracy of the phase analysis being better than 0.2 vol.%, the uncertainty in the volume fraction determination does not make any significant contribution to the overall uncertainty of the stress values. The finite strain measurement accuracy, $\sim$ 15 µstrain for diamond and 30-50 µstrain for SiC due to the neutron counting statistics is the major source of uncertainties of stress that are reported in Table 1. Additionally, some inaccuracy in d_0's and/or the single crystal elastic constants can result in somewhat bigger errors.

The experimental check for isotropy not only suggests isotropy of the microstructure and its properties but was used as an assumption in the stress analysis so that one stress component, the hydrostatic microstress (or pressure), fully characterizes the stress state.

Table 1. Experimentally determined phase composition and phase microstresses.

Sample ID	Weight fraction (diamond/SiC)	Volume fraction (diamond/SiC)	Stress in diamond [MPa]	Stress in SiC [MPa]
Cylindrical samples				
C#1	0.83/0.17	0.82/0.18	-339 ± 27	1524 ± 40
C#2	0.82/0.18	0.81/0.19	-323 ± 28	1334 ± 40
Prism samples				
T#1	0.88/0.12	0.88/0.12	-162 ± 35	1133 ± 46
T#2	0.92/0.08	0.91/0.09	-121 ± 33	1302 ± 54

Discussion

The results of the stress analysis on the samples made by the high pressure infiltration technique are consistent with other TSDCs [6], with hydrostatic microstresses in diamond phase reaching several hundreds of MPa in compression equilibrated by tensile microstresses in SiC with magnitudes up to GPa and higher.

Interpretation of the results can be made using a micromechanical model of the generalized particulate composite employing spherical inclusion theory. The stress analysis by means of the Hashin-Shtrikman bounds [11], rather than relying on one specific model, assumes isotropic and particulate microstructure and provides a range of possibilities for matrix-inclusion interaction. In this approach the hydrostatic microstress can be evaluated as a function of a temperature or external pressure drop. The calculations based on a temperature change from the sintering to room temperature are given Fig. 4 in comparison with the experimental data. It is evident that all samples generally match the lower HS bound ("HS-", diamond matrix, SiC inclusions) very well. This suggests an inter-connectivity of the diamond particles so the diamond phase topology can be considered as a continuous framework with islands of SiC.

MECA SENS 2017   Materials Research Forum LLC
Materials Research Proceedings **4** (2018) 65-70   doi: http://dx.doi.org/10.21741/9781945291678-10

With such good agreement between experimental and calculated stresses, some implications can be deduced.

(1) There is no pressure-drop-generated component in the overall stress state. If there was such a contribution present with pressure drop of $\Delta P = 9$ GPa, then the overall expected stresses would be significantly lower, e.g some -40 MPa rather than the -340 MPa experimentally determined for the diamond phase in the cylindrical samples. Thus only thermally generated stresses appear through the production pathway when the external pressure is released immediately, while temperature is reduced to ambient temperature at a slower rate. The same conclusion was made through surface stress analysis by means of Raman spectroscopy of the TSDC samples produced by the same technique [12] - the magnitude of residual stress is primarily dependent on temperature conditions, not pressure change.

(2) In the same way it can be deduced that there is no (or at least no significant) microcracking present in the samples, otherwise a significant stress relaxation with significantly lower absolute values of the microstresses would be detected. However, some speculations can be made regarding the triangular-shaped samples, where some stress relaxation can be observed, approximately 80 % from the expected values. The fact that the grain size of the diamond powder was smaller for these samples can be reflected in higher probability of the sealing effect with formation of isolated micro-pores. This can be responsible for the elastic weakening of the composite and can contribute to some stress relaxation.

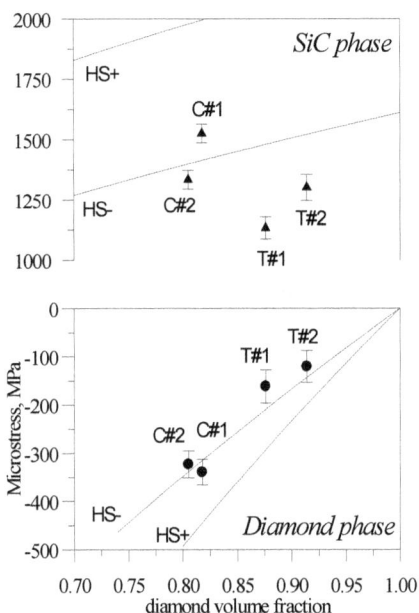

Fig. 4: Microstress in diamond and SiC phases for the TSDC samples: experimentally measured values (symbols) are plotted against predicted Hashin-Shtrikman bounds (lines).

Conclusions

Two types of TSDC samples (diamond-SiC composite), produced by the silicon infiltration technique at HTHP conditions, were analyzed by means of neutron diffraction. Phase composition was obtained through the full diffraction pattern analysis to assist in the residual stress analysis. Neutron diffraction strain analysis was done in a single peak experiment and microstresses (phase incompatibility) were determined in the two phases, compressive in the diamond phase and tensile in the SiC phase.

The thermal nature of the experimentally determined microstress was deduced through the micromechanical analysis of the composite. The results suggest that almost all thermally generated microstresses are preserved in the samples. No significant sign of microcracking and stress relaxation was found, thus confirming the micromechanical integrity of the samples and suggesting good quality and performance of the TSDC samples.

MECA SENS 2017   Materials Research Forum LLC
Materials Research Proceedings **4** (2018) 65-70   doi: http://dx.doi.org/10.21741/9781945291678-10

References

[1] J.N. Boland and X.S. Li, Microstructural characterisation and wear behaviour of diamond composite materials, Materials, 3 (2010) 1390-1419. https://doi.org/10.3390/ma3021390

[2] A.E. Ringwood, Diamond compacts and process for making same. Patent 4948388, (1989).

[3] K. Mlungwane, M. Herrmann and I. Sigalas, The low-pressure infiltration of diamond by silicon to form diamond–silicon carbide composites, J. Eur. Ceram. Soc. 28 (2008) 321-326. https://doi.org/10.1016/j.jeurceramsoc.2007.06.010

[4] C. Zhu, J. Lang and N. Ma, Preparation of Si–diamond–SiC composites by in-situ reactive sintering and their thermal properties, Ceram. Int. 38 (2012) 6131-6136. https://doi.org/10.1016/j.ceramint.2012.04.062

[5] G. Voronin, T. Zerda, J. Gubicza, T. Ungár and S. Dub, Properties of nanostructured diamond-silicon carbide composites sintered by high pressure infiltration technique, J. Mater. Res. 19 (2004) 2703-2707. https://doi.org/10.1557/JMR.2004.0345

[6] V. Luzin, J. Boland, M. Avdeev and X. Li, Characterization of Thermally Stable Diamond Composite Material, Mater. Sci. Forum 777 (2014) 165-170. https://doi.org/10.4028/www.scientific.net/MSF.777.165

[7] A.J. Studer, M.E. Hagen and T.J. Noakes, Wombat: The high-intensity powder diffractometer at the OPAL reactor, Physica B: Condensed Matter 385 (2006) 1013-1015. https://doi.org/10.1016/j.physb.2006.05.323

[8] B. Toby, EXPGUI, a graphical user interface for GSAS, J. Appl. Crys. 34 (2001) 210-213. https://doi.org/10.1107/S0021889801002242

[9] O. Kirstein, V. Luzin and U. Garbe, The Strain-Scanning Diffractometer Kowari, Neutron News 20 (2009) 34-36.

[10] K.S. Alexandrov and T.V. Ryzhova, The elastic properties of crystals, Sov. Phys. Crys., 6 (1961) 228–252.

[11] Z. Hashin, The Elastic Moduli of Heterogeneous Materials, J. Appl. Mech. 29 (1962) 143-150.

[12] M. Wieligor and T. Zerda, Surface stress distribution in diamond crystals in diamond–silicon carbide composites, Diamond Relat. Mater. 17 (2008) 84-89. https://doi.org/10.1016/j.diamond.2007.10.035

Processing & Welding

MECA SENS 2017   Materials Research Forum LLC
Materials Research Proceedings 4 (2018) 73-78   doi: http://dx.doi.org/10.21741/9781945291678-11

Investigating the Residual Stress Distribution in Selective Laser Melting Produced Ti-6Al-4V using Neutron Diffraction

L.S. Anderson[1,a], A.M. Venter[2,b], B. Vrancken[3,c], D. Marais[2], J. van Humbeeck[3,d] and T.H. Becker[1,e*]

[1]Department of Mechanical and Mechatronic Engineering, Stellenbosch University, South Africa

[2]Research and Development Division, South African Nuclear Energy Corporation, South Africa

[3]Department of Materials Engineering, University of Leuven, Belgium

[a]landerson@sun.ac.za, [b]andrew.venter@necsa.co.za, [c]bey.vrancken@kuleuven.be, [d]jan.vanhumbeeck@kuleuven.be, [e]tbecker@sun.ac.za

Keywords: Neutron Diffraction, Selective Laser Melting, Residual Stress, Additive Manufacturing, Ti-6Al-4V

Abstract. The Selective Laser Melting (SLM) process makes rapid manufacture of both prototype and structural components possible for a variety of metals. However, high residual stresses are inherent to the building process and can pose a number of problems; including part distortion, cracking and a reduction in the components' mechanical strength and fatigue life. Exact stress distributions through the part volume are often not known as the methods commonly used to measure residual stresses are surface or near surface stresses. As such, the influence of build parameters, residual stresses on resultant SLM-produced part integrity is not understood. Neutron diffraction allows for the measurement of residual stress through the volume of a part using measured lattice strains, thereby providing a tool to gain insight into the SLM process and the formation of residual stresses. In this work, neutron diffraction was used to determine the distribution of residual stress in a set of rectangular SLM-produced Ti-6Al-4V samples. Results show that an increase in layer thickness reduces the stress gradients in the part. There is also evidence that changing the exposure strategy can prevent stresses from developing along a preferential axis, making a more homogeneous stress field.

Introduction

Ti-6Al-4V is an α/β titanium-based alloy that contains, by weight, 4 % Al and 6 % V as its alloying elements [1]. Ti-6Al-4V is used extensively in the medical industry as well as in the aerospace industry in high strength-to-weight ratio components [2]. In recent years it has become a material of interest for use in Additive Manufacturing (AM) processes, in particular Selective Laser Melting (SLM) [1]. SLM is categorised as a powder bed fusion AM technology that selectively melts powder into a solid three-dimensional part using a high-powered laser source. SLM is capable of producing highly dense, net shape components [3]. As shown in Fig. 1, thin layers of atomized fine metal powder are evenly distributed using a coating mechanism from a powder hopper onto a baseplate that move along the vertical (z) axis. This takes place inside a chamber containing a tightly controlled atmosphere of inert gas, commonly argon. Once each layer has been distributed, each slice of the part geometry is fused by selectively melting the powder using a high-power laser beam, typically between 20 W and 1 kW [4]. The laser beam is directed (in the x and y axis directions) with two high frequency scanning mirrors. The laser energy is intense enough to permit full melting of the particles to form solid metal [5]. The process is repeated layer by layer until the part is complete.

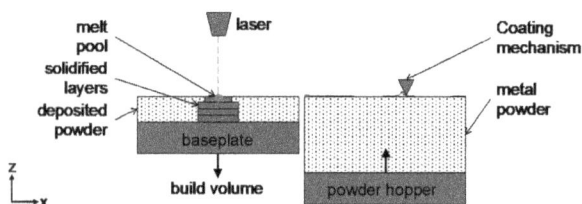

Fig 1. SLM process that selectively melts powder into a solid three-dimensional part using a high-powered laser source. Note the co-ordinate system where the x and y (out of the page) directions are parallel to the baseplate and the z direction is the build direction.

It is this melting and solidification cycle that results in the formation of residual stresses [6]. The primary mechanism is driven by a temperature gradient mechanism, which is a result of rapid, localized, heating that occurs at the impingement point of the laser on the material [7]. Large thermal gradients form due to the slow conduction of heat away from the melt pool, which result in a mismatch in the thermal expansion experienced by the molten material and the solidified material surrounding it [7]. The thermal expansion of the molten material is thus constrained by the solidified material surrounding it, resulting in compressive stress in the solid material. If the expansion is sufficient, the compressive stress in the constraining solid material exceeds its yield strength, causing it to deform plastically. Upon cooling of the molten region, a secondary mechanism occurs, whereby the thermal contraction of the molten region during the state change from liquid to solid is constrained by the plastically deformed solid material surrounding it. This constraint induces tensile stress in the newly solidified region and as a result the top build layer tends to develop a net tensile stress. A number of techniques, including X-ray diffraction, hole drilling and the contour method have been used to determine the influence of build parameters on residual stress in SLM Ti-6Al-4V, however, no investigation has been performed on the influence of build parameters on the stress distribution through the full volume of an SLM-produced Ti-6Al-4V part.

This study is aimed at investigating the through-volume residual stress distribution in a set of SLM-produced Ti-6Al-4V samples using neutron diffraction (ND). A brief introduction and background to the work is given first, followed by the details of the ND based stress measurements. The results obtained, and their relevance are presented and discussed.

Materials and methodology
The Ti-6Al-4V powder was acquired from 3D Systems Layerwise. A size distribution analysis showed that the particles ranged in size from 5 µm to 50 µm, with a median size of $d_{50} = 34.43$ µm. Rectangular samples with dimensions 20 x 20 x 10 mm^3 in x, y and z directions (as shown in Fig. 2(a)) were built at KU Leuven's Department of Metallurgy and Materials Engineering using an in-house developed SLM machine. The coordinate system used for the production of the samples is the same as that used in ASTM F2921, where the layers are deposited in the x-y plane and the sample is grown along the z-axis. All samples were wire cut from the base plate after manufacture, before residual stress measurements commenced.

The test matrix consisted of four samples; two layer thicknesses of 30 and 90 µm, and three exposure strategies, where the laser path would follow a parallel ([0]), perpendicular ([0/90]) and with a 30° ([0/30/60/...]) rotation to the previous layers laser path and are summarised in Table 1.

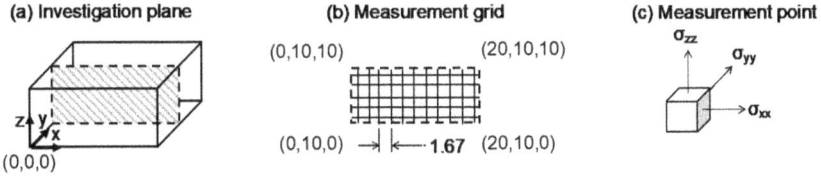

(a) Investigation plane   **(b) Measurement grid**   **(c) Measurement point**

(0,10,10)   (20,10,10)

(0,10,0) →| |← 1.67 (20,10,0)

Fig. 2 Depiction of the ND measurements points taken: (a) investigation plane, (b) 11x5
measurement grid of 1.67 mm spacing (in x and z) and (c) calculated stress tensor at a specific
grid point. All dimensions given in mm.

Table 1. Common and sample specific SLM build parameters used.

Common parameters		
Laser power [W]	Laser speed [mm/min]	Hatch spacing [µm]
250	1800	75
Sample-specific parameters		
Sample name	Exposure strategy[1] [°]	Layer thickness [µm]
1-30	[0]	30
2-30	[0/90]	30
2-90	[0/90]	90
3-30	[0/30/60…]	30

Archimedes test according to ASTM of the samples revealed an average density of minimum
of 99.42 % (sample 2-90) and a maximum of 99.58 % (sample 3-30).

ND measurements were performed at the South African Nuclear Energy Corporation (Necsa)
using the MPISI neutron strain scanning instrument. Measurements were taken through the centre
of the sample, along an investigation plane (normal to the y-axis in the x and z direction as shown
in Fig. 2a). This investigation plane was discretized into a 5 x 11 grid of measurement points
approximately 1.67 mm apart (as shown in Fig. 2b). Strain measured for the three orthogonal
directions matching the sample coordinate system were taken at each grid point using the
parameters detailed in Table 2 to provide normal stress measurements in σ_{xx} and σ_{yy}, as shown in
Fig. 2c. This measurement strategy was chosen on the assumption that the stress distribution
would be approximately symmetrical about the x-z plane due to the symmetry of the exposure
strategies and sample geometry. A 2 x 2 x 2 mm^3 ND analysis region was chosen to allow
sufficient accuracy in ND measurements, as well as sufficient spatial resolution.

Table 2. Parameters used for ND strain mapping.

analysis region [mm^3]	Psi tilt [°]	Bragg angle [°]	Miller indices of planes [hkl]	Average scan time per sample [hrs]
2 x 2 x 2	90	77.6	(103)	202

The zero-stress lattice spacing was determined for each sample by assuming that the normal
stress component in the z-direction (σ_{zz}) is zero. This was based on initial tests that indicated that
the stress in the normal z-direction is approximately zero i.e. that the material was in a biaxial
state of stress. The reference spacing d_0 was calculated using Eq. 1, based on which the σ_{xx} and
σ_{yy} could be computed using Eq. 2 [8].

$$d_0 = \frac{d_z(S_2 + 2S_1)}{S_1} - \frac{S_1(d_x + d_y)}{S_2},$$ (1)

In Eq. 1, d_0 is the calculated stress-free reference spacing, S_1 and S_2 are the diffraction elastic constants for {103} planes, and d_x, d_y and d_z are the measured plane spacings along the x, y, and z directions, respectively. In Eq. 2, σ_{ii} represents the normal stress components for σ_{xx} and σ_{yy} in units of MPa [8].

$$\sigma_{ii} = \frac{d_i - d_z}{d_0} * \frac{1 \cdot 10^6}{S_2}, \quad i = (x, y)$$ (2)

It is noted that in order to achieve a full stress tensor, six independent directions are needed to obtain the six stress components. This would result in testing durations becoming impractically long. The current setup of two stress directions and 55 sampling points requires about 10 days per sample. The average measurement error was 26.2 MPa for all samples. Respective stress gradients between adjacent points were calculated by numerically differentiating the obtained stress components and relative positions using central differencing scheme. This was done using Matlab's (ver. 2017a) gradient function.

Results and discussion

Stress distribution: Stress contour plots of σ_{xx} and σ_{yy} are shown for sample 2-30 in Fig. 3a. Stress values indicate an approximately parabolic shaped stress distribution in x and z. Furthermore, a symmetrical stress distribution is observed, as well as symmetry between σ_{xx} and σ_{yy}. Sample 3-30 showed a similar symmetry (not shown), however, sample 1-30 showed significantly lower stresses in the σ_{yy} direction, i.e. perpendicular to the laser path (not shown). This is due to the contraction of the heated material being more severely constrained in the laser path direction. Exposure strategies [0/90] and [0, 30, 60, ...], for which the laser path direction is alternated between each layer, tend to provide an equal stress state.

Influence of layer thickness: The influence of layer thickness is observed by comparing samples 2-30 and 2-90. The measured stresses are shown in Fig. 3b on the left at three build heights and respective stress gradients on the right. The comparison shows that a thinner layer thickness exhibits higher stress magnitudes (in both tension and compression), as well as steeper stress gradients. This result is consistent with previous research that showed that both the stress magnitude and gradient are a function of the number of build layers [7]. The data suggests that a larger layer thickness can offer advantages, from a residual stress point-of-view, by allowing for lower tensile and compressive stresses at the sample surface and centre respectively.

Influence of exposure strategy: The influence of the exposure strategy is considered in Fig. 3c by comparing samples 130, 230 and 330. Shown on the left are respective stress values at a build height of z = 5 mm and on the right the stress biaxiallity ratio, defined as σ_{xx}/σ_{yy}. Note that a value near one indicates equal stresses in σ_{xx} and σ_{yy}, whereas a value near 0.3 (Poisson's ratio) indicates a uniaxial stress state. The data shows a clear effect of the exposure strategy and stress homogeneity. Literature suggests that the exposure strategy influences the directionality of the stress distribution in the x-y plane. This directionality of stress is most evident when comparing sample 1-30 to sample 3-30, which shows a considerable variation in stress between the sample centre and surface. Both exposure strategies for sample 2-30 and 3-30 tilt the laser path and thereby the principal axes, resulting in a more homogeneous stress field. This is evident in the right figure (Fig. 3c) where sample 2-30 and 3-30 show a near biaxial stress state in the middle of the sample.

MECA SENS 2017   Materials Research Forum LLC
Materials Research Proceedings **4** (2018) 73-78   doi: http://dx.doi.org/10.21741/9781945291678-11

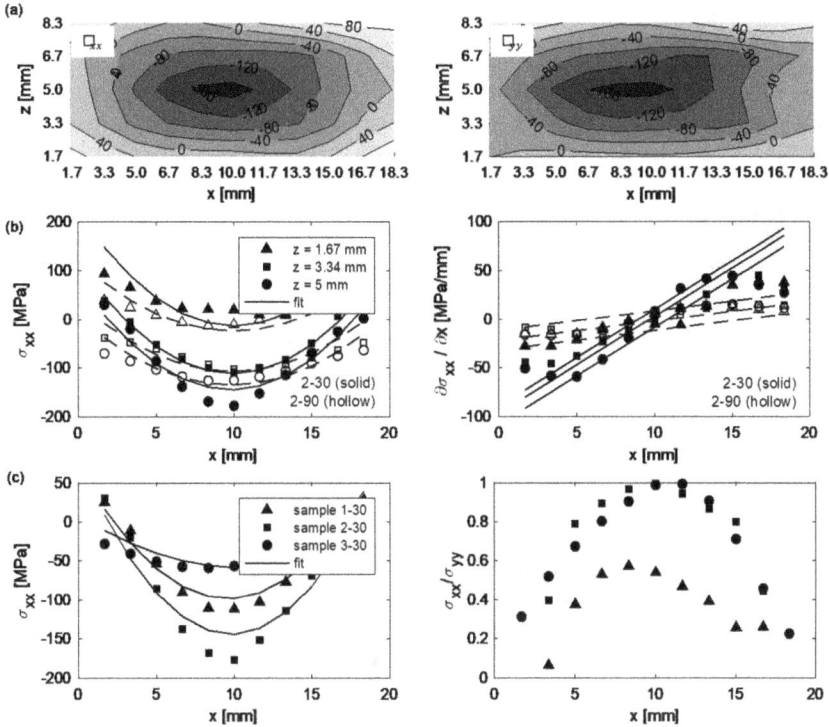

Fig. 3 (a) residual stresses for 2-30 in σ_{xx} (left) and σ_{yy} (right). (b) comparing stress values (left) and gradients (right) between a 30 and 90 µm layer thickness. (c) comparing stress values (left) and biaxiallity ratio (right) between the [0], [0/90] and [0, 30, 60, ...] exposure strategies.

Measurement discussion points: The use of ND for stress measurements in SLM produced Ti-6Al-4V uses analysis regions that are large compared to the build layers. For example, a sample built with 30 µm layer thickness results in 67 layers per analysis region. This means that the stresses are averaged over a large number of layers and as such the results obtained indicate an average stress value and not a maximum or minimum value.

Studies have shown that residual stresses decrease in magnitude by up to 80 % in the first 1.5 mm from the surface of SLM-produced parts [9]. Since this is less than the size of the analysis region, peak stresses cannot be obtained. Furthermore, high systematic errors that can occur if an analysis region is defined so that it protrudes from the sample and therefore surface and near-surface stress measurements cannot be performed. This loss of stress information is of vital importance as it is this near surface stress which most greatly affects the performance of SLM components.

One of the main limiting factors in this investigation was scan times. As indicated in Table 2, a single sample required 10 days of testing. Ideally, full mapping of the stress values in the samples should consider the full stress tensor consisting of 6 stress components [8]. This was not feasible, and the study assumed zero stress state in σ_{zz} based to initial testing undertaken. This stress state assumption is not a true representation of the state of stress within the samples, meaning that the use of Eq. 1 and Eq. 2 may not provide the correct stress values. However, the results do provide insight into the dependence of layer thickness and exposure strategy on the stress distribution. It is argued that the study provides sufficient evidence to motivate for exposure strategies that significantly alter the laser path between layers, as well as for larger layer thicknesses.

Conclusion

This study measured the bulk stress distribution in SLM-produced Ti-6Al-4V samples using ND. The study considered layer thickness and exposure strategy of identically sized rectangular samples. The following observations were made:

- Approximately parabolic shaped stress distributions were observed for rectangular samples.
- Stress magnitudes and gradients can be seen as a function of build layer thickness. By increasing the layer thickness, the stress magnitudes and gradients are significantly reduced.
- The exposure strategy used has a direct influence on the homogeneity of the stress state. A unidirectional exposure strategy results in preferentially orientated stress in the direction of the scan vector. By increasing the number of scan vector orientations, the stress components become more homogeneous and a biaxial stress state is achieved.

References

[1] M. Simonelli, Microstructure Evolution and Mechanical Properties of Selective Laser Melted Ti-6Al-4V, Loughborough University, 2014.

[2] H. J. Rack and J. I. Qazi, Titanium alloys for biomedical applications, Mater. Sci. Eng. C, 26(8) (2006) 1269-1277. https://doi.org/10.1016/j.msec.2005.08.032

[3] B. Ferrar, L. Mullen, E. Jones, R. Stamp and C. J. Sutcliffe, Gas flow effects on selective laser melting (SLM) manufacturing performance, J. Mater. Process. Tech. 212(2) (2012) 355-364. https://doi.org/10.1016/j.jmatprotec.2011.09.020

[4] D. Herzog, V. Seyda, E. Wycisk and C. Emmelmann, Additive manufacturing of metals, Acta Mater. 171 (2016) 371-392. https://doi.org/10.1016/j.actamat.2016.07.019

[5] S. Bremen, W. Meiners and A. Diatlov, Selective Laser Melting A manufacturing technology for the future, Laser Tech. J. 9(2) (2012) 33-38. https://doi.org/10.1002/latj.201290018

[6] M. F. Zaeh and G. Branner, Investigations on residual stresses and deformations in selective laser melting, Prod. Eng. 4(1) (2009) 35-45. https://doi.org/10.1007/s11740-009-0192-y

[7] P. Mercelis and J.-P. Kruth, Residual stresses in selective laser sintering and selective laser melting, Rapid Prototyp. J. 12(5) (2006) 254-265.

[8] M. J. Park, H. N. Yang, D. Y. Jang, J. S. Kim and T. E. Jin, Residual stress measurement on welded specimen by neutron diffraction, J. Mater. Process. Technol. 155-156 (2004) 1171-1177. https://doi.org/10.1016/j.jmatprotec.2004.04.393

[9] C. Casavola, S. L. Campanelli and C. Pappalettere, Preliminary investigation on distribution of residual stress generated by the selective laser melting process, J. Strain Anal. Eng. Des. 44 (2009) 93-105. https://doi.org/10.1243/03093247JSA464

MECA SENS 2017
Materials Research Proceedings **4** (2018) 79-84

Materials Research Forum LLC
doi: http://dx.doi.org/10.21741/9781945291678-12

Residual Stress Determination of Ductile Cast Iron by means of Neutron Diffraction

F. Smith[1,a*], J. Markgraaff[1,b], D.Marais[2,c] and A.M. Venter[2,d]

[1]North-West University, School of Mechanical and Nuclear engineering

[2]Research and Development Division, Necsa SOC Limited, Pretoria, South Africa

[a]swysnwu@yahoo.com, [b]johan.markgraaff@nwu.co.za, [c]deon.marais@necsa.co.za, [d]andrew.venter@necsa.co.za

Keywords: Valves, Ductile Cast Iron, Residual Stress, Neutron Diffraction, MAGMASOFT®

Abstract. A study was undertaken to simulate the casting process, using simulation software, of a ductile iron casting (for use as a valve body) and in doing so to determine the order of residual stress and experimentally verify the simulation results. The simulation was carried out using MAGMASOFT® and the residual stress results verified using neutron diffraction. The measured residual stress results were found to compare favourably with the simulation predictions.

Introduction

Mechanical valves play an essential part in industrial processes and other related systems. South African manufacturers must compete with international counterparts to supply affordable high-quality valves to clients. Production costs of valves can be reduced by optimising the design to use fewer raw materials.

Mechanical valves are most commonly produced from ductile cast iron. During the casting process, residual stresses are introduced in the component. By controlling the residual stresses within mechanical components, premature failure can be prevented.

Computer aided programmes, such as FEMLAB 3.1i and other, can be used to simulate certain aspects of the casting process. MAGMASOFT® with the module MAGMAstress aims to simulate the casting process and also predicts the presence of residual stress [1].

Method

The neutron diffraction analysis technique was chosen as the experimental verification method due to its non-destructive nature and the ability to penetrate deep within the material.

Sample selection. Due to the geometric complexities associated with valves, trees consisting of simple cylindrical shaped branches were chosen for this investigation and verification exercise. The same material (EN-GJS-400) as used in ductile iron valves manufactured in South Africa, was used.

Simulation. The cast tree (Fig. 1) consists of cylindrical branches with a height of 150 mm and different base diameters (15-30 mm). The residual stresses within the branches were simulated with MAGMASOFT® using a Solidworks model. Additional simulations were then carried out using boundary conditions which resembled heat treated branches and machined branches respectively (Fig. 2).

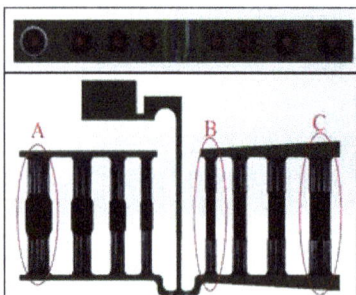

Fig. 1: (a) Sectional cut top view showing the 8 cylindrical branches' diameters and (b) front view of the ductile iron cast tree 3D model.

Fig. 2: Machined branches.

Simulations were conducted on the following branches with boundary conditions given in Table 1:
1. Branches simulated without runners (V2) (Fig. 4);
2. Branches simulated without runners subjected to a heat treatment process (V3) (Fig. 6); and
3. Branches simulated without runners subjected to machining (V4) (Fig. 9).

Table 1: MAGMASOFT® Simulation Information.

Pouring temp. [°C]	1400
Shakeout temp. [°C]	500
Pouring height [mm]	30
Mould sand	Furan Sand
Inlet size diameter [mm]	24
Pouring time [sec]	13
Machining size [mm]	Length: 120
	Diameter: 11.8 - 8 - 11.8 (see Fig. 2)
Heat treatment	1. Heated to 600 °C over 30 min
	2. Kept at 600 °C for 60 min
	3. Furnace cooled to 150 °C over 270 min
	4. Air cooled to room temperature

Table 2: Neutron diffraction measurement information.

Gauge volume [mm^3]		$2\times2\times2$
Slit positioning from centre of rotation [mm]		55
Wavelength [Å]		1.67
Diffraction peak measured		Fe (211)
2Θ detector angle [°]		90
Measurement time per position [sec]	V2	1200
	V3	3600
	V4	120

Verification. Three identical trees were cast using EN-GJS-400 ductile cast iron to replicate the simulated models. Three branches, A, B and C (Fig. 1), with the highest simulated residual stress values were cut from two of the trees.

The cavities MAGMASOFT® predicted were investigated by means of X-Ray Tomography using the facilities at the South African Nuclear Energy Corporation (Necsa) SOC Limited [2]. After model reconstruction (shown in Fig 3.), it was established that there were no cavities present in the cast sample. The neutron diffraction measurements could therefore be performed without correcting for spurious strains due to partial gauge volume illumination.

MECA SENS 2017   Materials Research Forum LLC
Materials Research Proceedings **4** (2018) 79-84   doi: http://dx.doi.org/10.21741/9781945291678-12

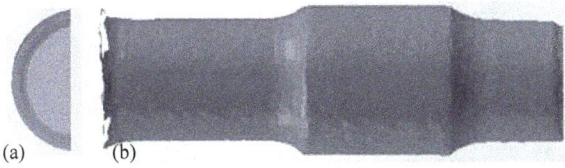

Fig 3: (a) Top and (b) front cut view of the reconstructed solid obtained by means of X-ray tomography.

(a)   (b)

The residual strains were then measured at the MPISI [3] angular dispersive neutron strain scanner situated at the SAFARI-1 research reactor of Necsa. The average d –spacing values measured of each branch, were used as the d_0 (reference) values for the measurements. Strains were then converted to residual stresses (Eq. 1) [4] along the measurement lines of the individual branches (Fig. 4, 6 and 9) and compared with the MAGMASOFT® simulation results. Information regarding the neutron diffraction measurements is summarized in Table 2.

$$\sigma_z = \frac{E}{(1+v)(1-2v)}[(1-v)\varepsilon_z + v(\varepsilon_x + \varepsilon_y)] \tag{1}$$

Results

The stress values were extracted of the z-direction stress components along the z-axis (vertical axis) direction (Fig. 4) as this was where MAGMASOFT® predicted the highest order of residual stress. The residual strain measurements were obtained at 30 coordinates, in comparison with 100 coordinates MAGMASOFT® used, along the z-axis direction.

Graphs show the difference between the MAGMASOFT® and neutron diffraction values for every simulation done. Comparison graphs of (V2) vs (V3) and (V2) vs (V4) are also shown. MAGMASOFT® and neutron diffraction results of only Branch A will be shown.

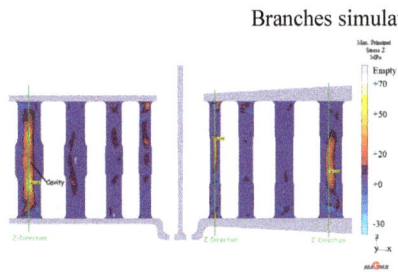

Branches simulated without runners (V2)

Fig. 4: MAGMASOFT® Z-component residual stress results – Simulation (V2) and measurement directions shown.

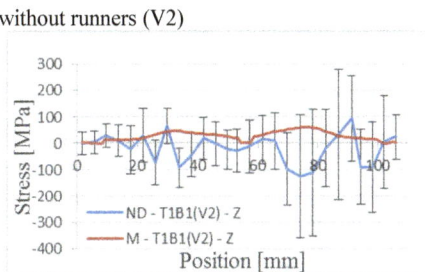

Fig. 5: MAGMASOFT® vs Neutron Diffraction Z-component residual stress results (V2).

MECA SENS 2017   Materials Research Forum LLC
Materials Research Proceedings **4** (2018) 79-84   doi: http://dx.doi.org/10.21741/9781945291678-12

Heat treated branches (V3)

Fig. 6: MAGMASOFT® Z-component residual stress results – Simulation (V3) heat treated.

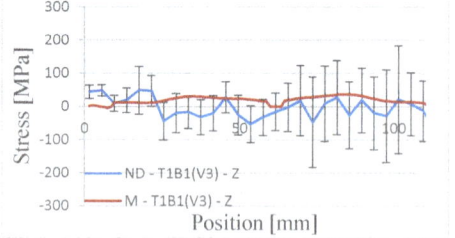

Fig. 7: MAGMASOFT® vs Neutron Diffraction Z-component residual stress results (V3) heat treated.

Heat Treated Branches: Stress vs Position

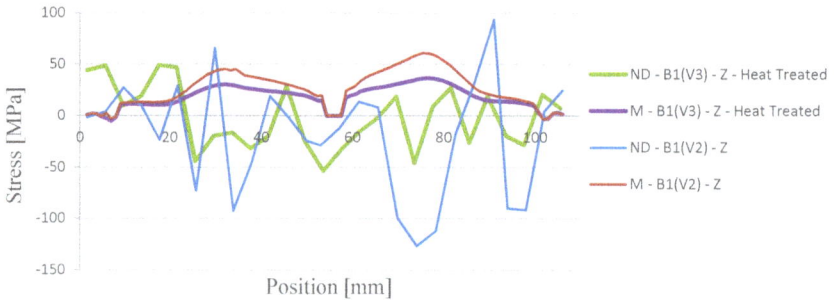

Fig. 8: Comparison of Z-component residual stress results between (V2) and (V3).

Machined branches (V4)

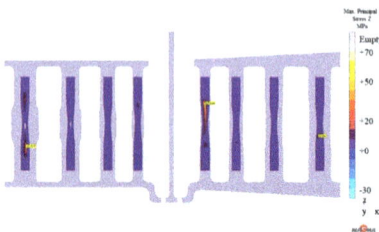

Fig. 9: MAGMASOFT® Z-component residual stress results – Simulation (V4) machined.

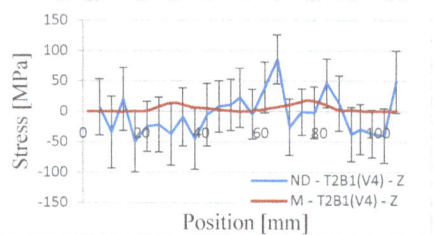

Fig. 10: MAGMASOFT® vs Neutron Diffraction Z-component residual stress results (V4) machined.

MECA SENS 2017
Materials Research Forum LLC
Materials Research Proceedings **4** (2018) 79-84
doi: http://dx.doi.org/10.21741/9781945291678-12

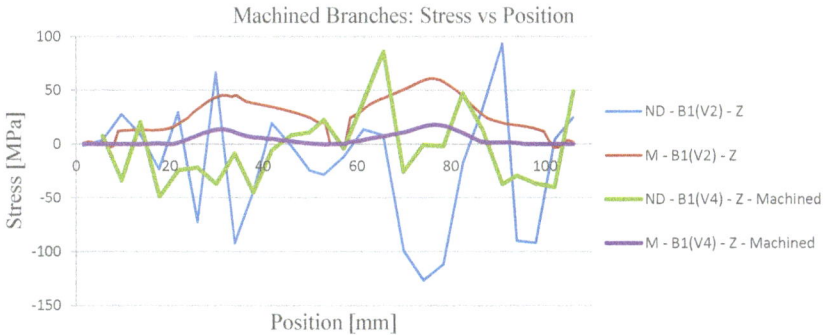

Fig. 11: Comparison of Z-component residual stress results between (V2) and (V4).

Discussion

Considering the X-ray tomography results, no cavities were observed, thus the neutron diffraction measurements could continue normally without any corrections.

With the data obtained trends between MAGMASOFT® and the neutron diffraction results (Figs. 5, 7 and 10) are apparent, even though the values do not match very well. Most of the values fall within the experimental error limit.

When analyzing the data, peaks can be observed in the neutron diffraction measurement results, whereas the MAGMASOFT® results have even curves. This is due to the measurements not having as much data points as the MAGMASOFT® simulation results.

When considering the heat treated and machined results (Figs. 8 and 11), both MAGMASOFT® and neutron diffraction predicted a decrease in residual stress.

Possible reasons for the inaccuracy of the residual stress results:

- The cope and the drag of the mould was not clamped and sealed correctly causing a leak through the mould forming a large flash, which possibly influenced the residual stress results.
- The material used for casting could have differed slightly in composition to that specified in the simulation.

Conclusion

The data collected suggests that MAGMASOFT® can be used to accurately simulate the residual stress in the casting process and more specifically simulate residual stress in valves as it uses conservative measures.

To fully verify MAGMASOFT®, samples should be made consisting of much higher residual stress which can be more accurately determined by means of neutron diffraction.

The neutron diffraction statistics could also be improved by means of increasing the measurement time per position, thus decreasing the experimental errors.

References

[1] Information on http://www.magmasoft.com/en/solutions/MAGMA_5.htm

[2] J.W. Hoffman and F.C. de Beer, Characteristics of the Micro-Focus X-ray Tomography Facility (MIXRAD) at Necsa in South Africa, 18th World Conference on Non-destructive Testing. (2001)

[3] A.M. Venter, P.R. van Heerden, D. Marais and J.C. Raaths, MPISI: The neutron strain scanner Materials Probe for Internal Strain Investigations at the SAFARI-1 research reactor, Physica B, In press. https://doi.org/10.1016/j.physb.2017.12.011

[4] G.A. Webster and R.C. Wimpory, Non-destructive measurement of residual stress by neutron diffraction, J. Mater. Process. Technol. 117 (2001) 395-399. https://doi.org/10.1016/S0924-0136(01)00802-0

MECA SENS 2017
Materials Research Proceedings 4 (2018) 85-90

Materials Research Forum LLC
doi: http://dx.doi.org/10.21741/9781945291678-13

Effect of Residual Stress Relaxation due to Sample Extraction on the Detectability of Hot Crack Networks in LTT Welds by means of µCT

F. Vollert[1,a*], J. Dixneit[2,b] and J. Gibmeier[1,c]

[1]Karlsruhe Institute of Technology (KIT), Institute of Applied Materials (IAM), Kaiserstr.12, 76131 Karlsruhe, Germany

[2]Bundesanstalt für Materialforschung und –prüfung (BAM), Unter den Eichen 87, 12205 Berlin, Germany

[a]florian.vollert@kit.edu, [b]jonny.dixneit@bam.de, [c]jens.gibmeier@kit.edu

Keywords: LTT Weld Filler Materials, µCT-analysis, Hot Cracks, Welding

Abstract. Investigations on weldability often deal with hot cracking as one of the most prevalent failure mechanisms during weld fabrication. The modified varestraint transvarestraint hot cracking test (MVT) is well known to assess the hot cracking susceptibility of materials [1, 2]. The shortcoming of this approach is that the information is only from the very near surface region which inhibits access to the characteristic of the hot crack network in the bulk. Here, we report about an alternative approach to monitor the entire 3D hot crack network after welding by means of microfocus X-ray computer tomography (µCT). However, to provide sufficient high spatial resolution small samples must be sectioned from the MVT-welded joint. The sampling is accompanied by local relaxation of the residual stress distributions that are induced by welding, which can have an impact on the crack volumes prior to the sampling. The studies were carried out to investigate the hot cracking susceptibility of low transformation temperature filler materials (LTT) [3, 4]. As high compression residual stresses up to -600 MPa in the area of the crack networks were determined by means of the contour method, stress relaxation caused by sectioning for µCT sample extraction can affect the detectability of the cracks later on. X-ray diffraction studies revealed surface residual stress relaxations up to about 400 MPa due to cutting. To investigate this effect, the specimens with hot cracks were subjected to a load test with known stress states. The results clearly show that local stress relaxations will have a strong impact on the volume images reconstructed from tomography analysis. This effect must be considered during hot crack assessment on basis of µCT data.

Introduction

Using low transformation temperature (LTT) weld filler materials is an innovative method to mitigate welding tensile residual stresses directly during weld fabrication by means of a delayed martensite transformation and thus, the fatigue life will be increased without time consuming or cost intensive post-weld treatments [5, 6]. However, LTT alloys are high alloyed filler materials and therefore may show high hot cracking susceptibilities dependent on the chemical composition. Measurements of the total crack length provided by µCT analysis in the MVT showed that the hot crack susceptibility of LTT alloys will be increased with increasing content of chrome and nickel, while the total crack length maximum will be in the weld depth [4]. As an appropriate approach to assess hot cracking, the standardised modified varestraint transvarestraint hot cracking test (MVT) was developed [1]. By means of this test different base or filler materials can easily be evaluated during welding while bending the specimen at the same time in the longitudinal or transversal directions to the weld line using defined bending rates.

MECA SENS 2017   Materials Research Forum LLC
Materials Research Proceedings **4** (2018) 85-90   doi: http://dx.doi.org/10.21741/9781945291678-13

Hereby tungsten inert gas (TIG) welding is used to simulate the weld fabrication by means of different heat inputs. After welding the total length of all detectable cracks at the specimen surface is added up. Depending on the bending strain applied, the hot cracking susceptibility of the tested material will be ranked into *'safe against hot cracks'*, *'increasing hot crack susceptibility'* and *'high hot crack susceptibility'*. However, MVT does not consider a crack network in the bulk. Hence, for accurate assessment of the hot crack susceptibility of material volumes information shall also be considered in addition. Therefore, microfocus X-ray computer tomography (µCT) was applied (Fig. 1a). This method allows describing the complete 3D hot crack network in regards to e.g. crack volume, crack length as a function of the distance to the surface (Fig. 1b) and crack orientation.

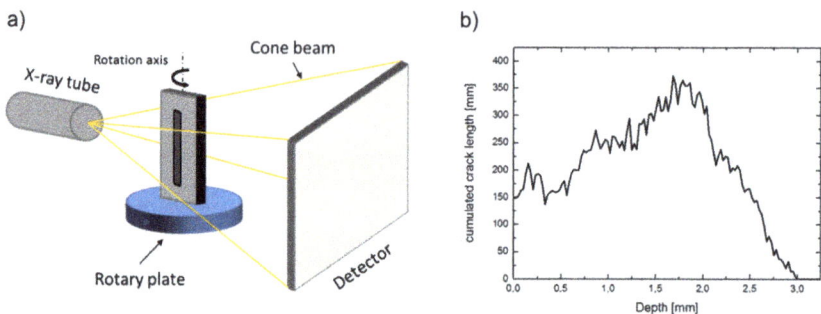

Fig. 1: Schematic illustration of the µCT-scan of the whole specimen to obtain the position of the hot crack network (a). As an example, the cumulated total crack length as a function of depth for the investigated specimen determined by µCT is plotted (b).

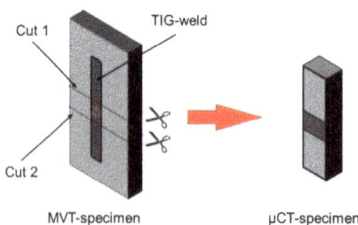

Fig. 2: Sample preparation for µCT studies with initial MVT-specimen (left) and µCT-specimen (right). Location of hot cracks are indicated in red.

To provide a sufficient high resolution and to reduce X-ray absorption of the specimen it is necessary to cut the investigated specimen into smaller cuboids (Fig. 2). Because of the cone beam used in lab CT analysis (see Fig. 1a) the image resolution becomes better the closer the specimen can be positioned to the X-ray tube, which requires tailored and adequate specimen dimensions. However, the cutting process leads to residual stress relaxation in the region of the crack network, which may alter the results. In this paper we investigate the effect of residual stress relaxations on the visibility of cracks in the volume images re-constructed from µCT-scans.

MECA SENS 2017   Materials Research Forum LLC
Materials Research Proceedings **4** (2018) 85-90   doi: http://dx.doi.org/10.21741/9781945291678-13

Experimental

Material and specimen geometry. The investigated samples were standardised MVT specimens with dimensions 100 x 40 x 10 mm^3. Prior to testing along the MVT welding direction a U-shaped groove with a depth of 5 mm and a width of 20 mm was milled into the blocks. The LTT alloy was deposited into the groove by manual gas metal arc welding using six layers while the low alloyed high strength steel S960Q was used as a substrate to reduce the amount of filler material. After welding the specimens were finished to the standardised MVT dimensions to account for the weld reinforcement. The chemical composition is shown in Table 1. During the MVT the LTT welds were re-melted by automated TIG-welding using a heat input of 7.5 kJcm^{-1}. Bending of the MVT specimens were executed during welding in the longitudinal direction to the weld line (varestraint-modus) using a bending radius of 125 mm (resulting surface strain of 4 %). A μCT-scan of the whole MVT-specimen was performed to obtain the position of the hot crack network (Fig. 1a).

Table 1: Chemical composition in wt.% of the pure LTT alloy and the base material determined by spectral analysis.

Material	Chemical composition in wt.%							
	C	Cr	Ni	Si	Mn	Mo	V	Fe
LTT weld	0.11	6.5	7.9	-	0.59	-	-	bal.
S960Q (base material)	0.18	0.8	2.0	0.5	1.6	0.6	0.1	bal.

Residual stress measurements. Residual stress measurements were performed using the contour method developed by M. Prime [e.g. 7]. Here, the investigated specimen was cut at a chosen measuring plane (49 mm distance to specimen edge) using electric discharge machining (EDM). Afterwards the resultant deformation caused by residual stress relaxation at the newly created surface was measured by a coordinate measuring machine (CCM). The measured deformations are transferred to a finite element (FE) model and a linear elastic stress analysis performed in order to calculate a residual stress map of the stress component normal to the cut surface. This method is a well-suited technique for the investigated specimens as it gives an entire 2D residual stress map of the whole weld joint. In this paper the longitudinal residual stress component of one MVT-specimen was determined. Therefore, a cut in the transverse direction to the weld line was carried out using EDM with a brass wire of diameter 0.3 mm. The specimen was clamped as close as possible to the cut to prevent sample movement. Afterwards the two cut surfaces were measured using the CCM type PRISMO navigator from Zeiss. A clamping styli with a ruby sphere (diameter = 0.67 mm) was used with measuring point spacing set to 0.1 μm. The deformation data of both cut surfaces were averaged and obvious outliers removed. The resultant displacements were transferred as boundary conditions to a FE model of the investigated specimen. With Abaqus® a fully elastic FE stress analysis using Young's modulus of 210 GPa and Poisson's ratio of 0.3 was performed. The results provide the initial residual stress state before cutting.

μCT load test. The load test was carried out using a load specimen, which was prepared out of the weld of an additionally tested MVT-specimen. The load specimen (Fig. 3 left) was cut out in the region of the cracks (determined by the μCT-scan before) using EDM. A load testing machine (Fig. 3 right) specifically developed for μCT-studies was used to investigate the effect of known stress states on the crack network. 9 load steps of compressive stress from 0 MPa up to -325 MPa were applied to the specimen.

MECA SENS 2017   Materials Research Forum LLC
Materials Research Proceedings 4 (2018) 85-90   doi: http://dx.doi.org/10.21741/9781945291678-13

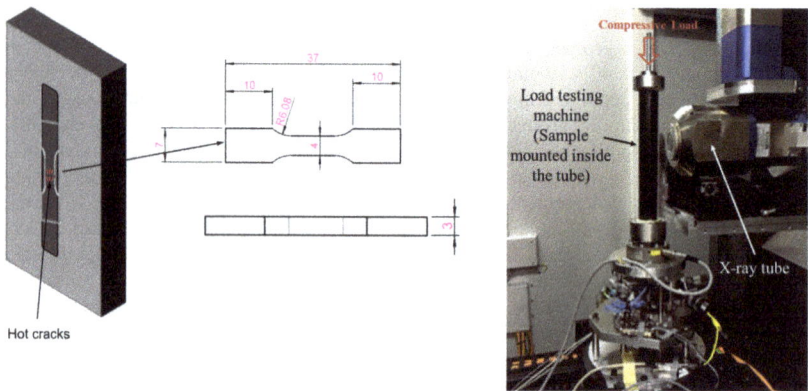

Fig. 3: Removal position of the load specimen for the load test (left). Load testing machine applied for the in-situ loading µCT-studies (right).

During each load step a µCT-scan was performed while keeping the load at a constant level. The µCT-scans were done using a µCT system type Y.CT.Precision (Yxlon ltd., Hamburg, Germany) with tube voltage and current set to 160 kV and 0.05 mA, respectively, providing a maximum resolution of about 1 µm voxel size. Here, to get maximal high resolution, the load testing machine is positioned as close as possible to the X-ray tube resulting in a voxel size of 25 µm. The 3D images of the crack network were segmented into crack or base material using an adaptive threshold algorithm [8]. An adapted watershed algorithm [9], using the cracks determined by the adaptive thresholding algorithm as seeds, was then performed. This segmentation approach with defined thresholds ensures comparability between the different load steps. The segmented volume images can then be used to describe the hot crack network in regard to e.g. total crack volume and crack length as a function of applied stress.

Results and Discussion

Figure 4 displays the residual stress map for the longitudinal component determined by the contour method where the sample was cut perpendicular to the weld line. Compressive residual stresses up to approximately -600 MPa are present in the re-melted LTT weld. The region of compressive residual stresses extends down to a depth of approximately 4 mm. Balancing tensile residual stresses up to 600 MPa enclose the area of compressive residual stresses. The µCT-scan of the whole MVT-specimen (prior to cutting the specimen for the load test) revealed that the complete hot crack network is located in the compressive residual stress area. X-ray diffraction studies revealed surface residual stress relaxation up to about 400 MPa due to the cutting process. It may therefore be assumed that residual stress relaxation occurs in the region of the hot cracks. Figure 5 shows the 3D hot crack network in the tensile specimen before load was applied. The evaluation of the total crack network based on µCT data showed a decrease of total crack volume with increasing compressive stress (Fig. 6a). To investigate the reason of this effect, studies on individual cracks were performed. These cracks (mean crack lengths approx. 250 µm) have been selected based on their orientation to the load direction. The focus was set on two cracks and their behaviours when compressive stress was applied. One crack is oriented at 0 degrees (load type: mode III) and the other at 90 degrees (load type: mode I) to the load direction (Fig. 5).

MECA SENS 2017
Materials Research Forum LLC
Materials Research Proceedings **4** (2018) 85-90
doi: http://dx.doi.org/10.21741/9781945291678-13

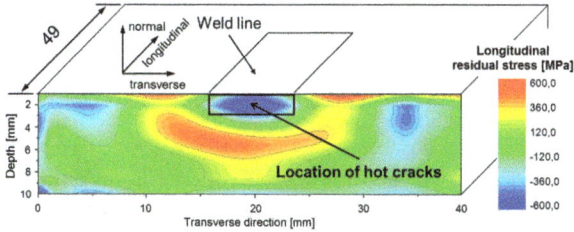

Fig. 4: Residual stress map of the longitudinal component determined by the contour method.

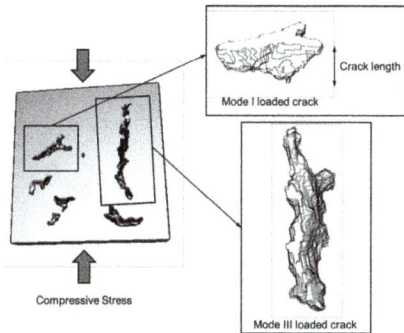

Fig. 5: 3D hot crack network to be found in the tensile specimen and the two investigated cracks (mode I and mode III loaded).

Analysis of the crack volume as a function of applied load leads to no measurable volume change for the mode III crack. However, the loaded mode I crack shows a decrease in crack volume with increasing compressive stress. A detailed observation reveals that the volume change is caused by the change in crack length (Fig. 6b). Apparently, increasing compressive stress leads to crack closure. The distance between opposing crack surfaces needs to be large enough to have an influence on the X-ray absorption and thus on the detectability in the reconstructed image. Therefore, the crack length seems to decrease in the μCT images, the higher the applied compressive stresses are.

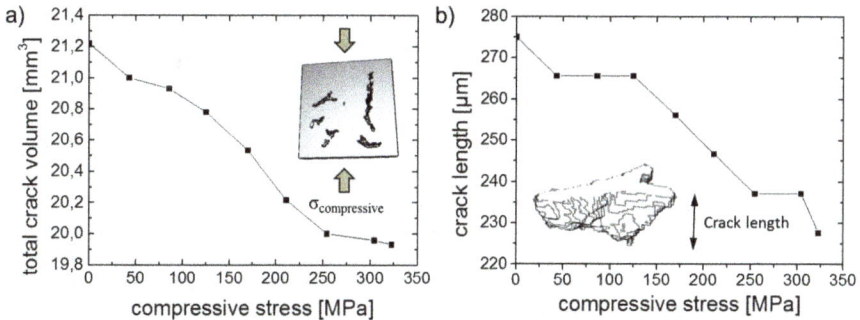

Fig. 6: Total crack volume as a function of the applied compressive stress (a) and crack length of the mode I loaded crack as a function of the applied compressive stress (b).

As a result, high compressive residual stresses in the region of the hot crack network can affect the detectability of the true crack length. Assuming that the actual crack length remains constant during sample preparation, in the case of compressive residual stresses in LTT-welds, cutting of the MVT-specimens and the resultant compressive residual stress relaxation may have a positive effect to determine the true total crack length using μCT.

Conclusion

The effect of residual stresses on the detectability of hot cracks using μCT was investigated. The following major statements can be derived:

- Compressive residual stresses that are present in LTT-welds of MVT test samples relax during sample preparation for μCT and affect the detectability of hot cracks.
- The total crack length seems to be underestimated in presence of compressive residual stresses in the weld as the crack surfaces are pressed together and therefore the entire crack length is not determinable in reconstructed images.
- Compression tests revealed that the detectability of the crack length of cracks oriented 90 degrees to the load direction is strongly affected, while cracks oriented 0 degrees to the load direction show no significant effect.
- Furthermore, the μCT studies showed that providing the 3D crack network results are a much better basis for the assessment of the hot crack susceptibility in contrast to the conventional MVT test, which only focus on surface information.

Acknowledgement

Financial support by the German Research Foundation (DFG) in the projects GI376/8-1 is gratefully acknowledged.

References

[1] DIN EN ISO 17641: Destructive tests on welds in metallic materials—hot cracking tests for weldments, part 1, 3—arc welding processes (2005)

[2] Th. Kannengiesser and Th. Boellinghaus, Hot cracking tests-an overview of present technologies and applications, Weld World 58 (2014) 397-421. https://doi.org/10.1007/s40194-014-0126-y

[3] A. Ohta, O. Watanabe, K. Matsuoka, C. Siga, S. Nishijima, Y. Maeda, N. Suzuki and T. Kubo, Fatigue strength improvement by using newly developed low transformation temperature welding material, Welding in the World 43 (1999) 38–42.

[4] Th. Kannengiesser, M. Rethmeier, P.D. Portella, U. Ewert and B. Redmer, Assessment of hot cracking behaviour in welds, Int. J. Mater. Res. 102(8) (2011) 1-6. https://doi.org/10.3139/146.110545

[5] E. Harati, L. Karlsson, L.-E. Svensson and K. Dalaei, Applicability of low transformation temperature welding consumables to increase fatigue strength of welded high strength steels, Int. J. Fatigue 97 (2017) 39-47. https://doi.org/10.1016/j.ijfatigue.2016.12.007

[6] J. Gibmeier, E. Obelode, J. Altenkirch, A. Kromm and Th. Kannengießer, Residual stress in steel fusion welds joined using low transformation temperature (LTT) filler material. Mater. Sci. Forum 768-769 (2014) 620-627. https://doi.org/10.4028/www.scientific.net/MSF.768-769.620

[7] M.B. Prime and A.T. Dewald, Chapter 5 The contour method, in: G. S. Schajer (Ed.), Practical Residual Stress Measurement Methods, Wiley-Blackwell, 2013, pp. 109-138. https://doi.org/10.1002/9781118402832.ch5

[8] D. Bradley and G. Roth, Adaptive thresholding using the integral image, J. Graph. Tools 12 (2007) 13-21. https://doi.org/10.1080/2151237X.2007.10129236

[9] S. Beucher and C. Lantuejoul, Use of watersheds in contour detection, workshop on image processing, real-time edge and motion detection (1979)

MECA SENS 2017
Materials Research Proceedings **4** (2018) 91-96

Materials Research Forum LLC
doi: http://dx.doi.org/10.21741/9781945291678-14

Fast Temporal and Spatial Resolved Stress Analysis at Laser Surface Line Hardening of Steel AISI 4140

D. Kiefer[1,a*], J. Gibmeier[1,b] and F. Beckmann[2,c]

[1]Karlsruhe Institute of Technology (KIT), Institute for Applied Materials (IAM-WK), Engelbert-Arnold-Str. 4, 76131 Karlsruhe, Germany

[2]Institute of Materials Research, Helmholtz-Zentrum Geesthacht (HZG), Max-Planck-Str. 1, 21502 Geesthacht, Germany

[a]Dominik.Kiefer@kit.edu, [b]Jens.Gibmeier@kit.edu, [c]Felix.Beckmann@hzg.de

Keywords: Laser Hardening, X-ray Diffraction, Synchrotron Radiation, Real Time Stress Analysis

Abstract. Local and temporal strain and stress evolution is recorded by synchrotron X-ray diffraction during laser line hardening of SAE 4140 steel in the quenched and tempered states at different measuring positions with respect to the process zone. The in-situ diffraction experiments were performed at beamline P05@Petra III at DESY, Hamburg (Germany). The steel samples were line hardened using a 4 kW high-power diode laser (HPDL) unit at a constant laser feed of 800 mm/min. Using a specially designed process chamber that incorporates symmetrically attached fast silicon micro-strip line detectors, stress analysis using the $\sin^2\psi$-method in single-exposure mode, enabled measuring rates at 20 Hz. As a result of the temporal and spatial resolved analyses, the elastic strains were separated from the thermal strains.

Introduction

In the last decades, with the development of high power diode lasers (HPDL), laser surface hardening gained increased interest for the provision of localized fatigue and wear resistant surface regions of technical components. The process is characterized by localised heat input using a laser beam, followed by self-quenching. Using fiber coupled laser optics the hardening process is rather flexible. Hence, typical applications for structural components are, inter alia, cutting edges, turbine blades or forging matrices that are locally tailored hardened martensite to the application requirements. By this means beneficial residual stress (RS) states can be locally induced in the near surface region that impedes e.g. failure through crack initiation. The advantages over competing surface hardening processes are high automation capability, fast processing and minimal distortion due to the local heat input. However, process prediction is very complex and mainly based on case studies in the final state for the particular material [1-4]. We demonstrated in previous studies that in-situ X-ray diffraction experiments are a suitable tool to investigate fast running thermal and thermo-chemical processes such as laser surface hardening. Here, quick time resolutions require high X-ray photon flux, which can only be provided by synchrotron sources. Results of in-situ diffraction studies during processing help deepen the process understanding and can be used to improve and validate process simulation and thereby allow accurate process predictions. The adaption of the single exposure technique [5] into an experimental setup for laser surface hardening as presented in [6, 7] allows for the real-time monitoring of phase transitions and strain evolutions. Here, a specially designed process chamber was commissioned with measuring and evaluation strategies established that allows for the separation of thermal and elastic strains for each exposure. Finally the stresses were calculated according to the well-known $\sin^2\psi$-method [5]. This set-up was developed further. Earlier work [6] started exclusively with spot hardening experiments. We equipped the process chamber with a motorised tilt holder for the laser optics that allows for defined laser line hardening experiments. Additionally we established new laser optics with an in-line single color pyrometer that provides fully temperature-controlled processing during laser line hardening. The process chamber, described in [6], was upgraded with more suitable linear motors to provide laser feed speeds on a technically relevant scale. Here, we report about new results that were recorded using this upgraded set-up and about the achievements reached. The results of temporal and spatial resolved phase-, strain- and stress-evolution are discussed.

Experimental

Material and sample preparation: For the laser line hardening experiments cuboid samples $(80 \times 50 \times 15 \text{ mm}^3)$ made from AISI 4140 steel in a quenched and tempered state were mechanically ground for improved surface smoothness to increase laser absorption and to guarantee a consistent surface quality. Subsequent, the samples were subjected to a stress relief heat treatment at 510°C for 90 min under inert gas atmosphere.

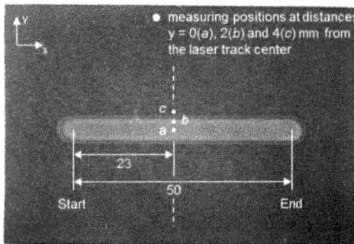

Fig. 1: Image of laser line hardened sample with the different measuring positions, which were set at 23 mm from the start for the in-situ experiment.

Experimental setup and implementation: In-situ X-ray diffraction laser line hardening experiments were carried out at the beamline P05@PETRA III operated by the Helmholtz Center Geesthacht (HZG) at DESY in Hamburg, Germany. Synchrotron radiation was provided by a double crystal monochromator and set to E = 10.899 keV (λ = 1.1384 Å). Laser feed speed was set to 800 mm/min from the starting position at a temperature T_{max} = 1150 °C. The laser system comprised a fiber coupled 4 kW high power diode laser unit type LDM 4000-100 in combination with Gaussian focusing optics with a nominal focal point diameter of 5.8 mm: Both from Laserline GmbH, Mühlheim-Kärlich (Germany). Tracks with a length of 50 mm and a nominal width of approx. 5 mm were

laser hardened. The X-ray synchrotron investigations were done with a double cross slit that adjusted the X-ray spot size to about $1 \times 1 \text{ mm}^2$. Measurements were taken at three different positions *a*, *b* and *c* that corresponded to distances 0, 2 and 4 mm to the center of the laser track axis as illustrated in Fig. 1. The atmosphere inside the process chamber is set to a low He overpressure to avoid oxide scale formation during the process. The sample is pre-tilted, with respect to the primary beam axis, by an angle χ = -35°. Key components of the setup are fast micro-strip detectors (MYTHEN-1K, Dectris Ltd.), which are radially (r = 200 mm) arranged around the process chamber in a manner that correspond to the single exposure technique described in [4]. The measuring frequency was set to 20 Hz ($t_{exposure}$ = 50 ms). At the given wavelength both detectors cover a 2θ range of about 140° - 157°. A scheme of the experimental setup is shown in Fig. 2. The setup allows for the simultaneous measurement of a diffraction peak *hkl* from the identical diffraction cone under two different tilt angles ψ_1 and ψ_2. By definition, ψ is the angle between the sample surface normal P_3 and the diffraction vector $N_i^{\{hkl\}}$.

Data processing and analysis: For the detector calibration, powder samples of LaB$_6$ and α-Fe were used. Prior to peak fitting an absorption correction and a linear background subtraction were performed. The diffraction peaks are fitted using a Pseudo-Voigt function. Error bars were calculated on the basis of a 95 % confidence interval for the peak fits. At the chosen synchrotron radiation wavelength the $\{422\}$ α-Fe ($2\theta_0$ = 153.206°) and the $\{600\}$ γ-Fe ($2\theta_0$ = 138.790°) peaks

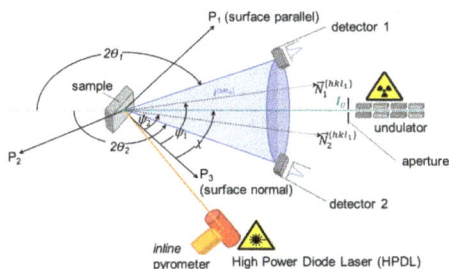

Fig. 2: Scheme of the experimental in-situ setup with diffraction cone and indicated angles.

were measured. Thermal strains result in a vertical shift of the 2θ-$\sin^2\psi$ line plot due to the influence on the hydrostatic part of the stress tensor. On the other hand, elastic strains result in a change of slope. A separation of thermal and elastic strains was performed as described in [6] and the deviatoric stress evolution during the process calculated. Here we focus on the determination of the strain / stress components transverse to the laser track (= transverse to the feed direction). The stress independent lattice directions were calculated from the phase-specific, temperature dependent macroscopic Young's moduli E and Poisson ratio v.

Post-process investigations: The results of the in-situ stress analyses were compared to high spatially resolved RS lab analyses according to the classical $\sin^2\psi$ -method. 41 positions were measured over the laser treated line with an increment step width $\Delta y = 0.25$ mm. The measurements were performed using a ψ-diffractometer and V-filtered CrK_α- radiation. Here, the {211} α-Fe diffraction line ($2\theta_0 = 156.394°$) was measured at 13 ψ angles for -60° < ψ < 60°. The primary beam was collimated using a 100 µm focusing polycapillary optics. On the secondary side a 4 mm symmetrical slit was used in front of the scintillation counter.

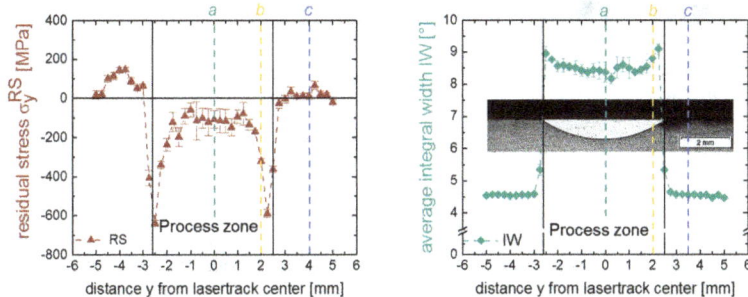

Fig 3: Results of high resolution RS (transverse component) lab measurements (ex-situ) across the laser track (left) and corresponding average integral peak width of the diffraction lines (right). The in-situ (synchrotron) measuring positions a, b and c is represented by the dashed lines on the two graphs. Also shown (right) is a cross-sectional micrograph of the hardened zone.

Results and discussion

A metallographically prepared cross section of a laser line hardened sample shows an ideally sectorial martensitic hardened zone with a maximum width of 5.1 mm at the surface and a maximum depth of 0.81 mm in the center (see Fig. 3 background). Using a confocal 3D microscope, the surface topography was analyzed post-process showing a peak in the center of the laser track of about 4.3 µm in height compared to the not hardened surface. The results of the high spatially resolved lab measurements (ex-situ study) are plotted in Fig. 3 (left: RS | right: integral peak widths of the diffraction lines).

Transverse residual stresses at the surface show a characteristic W-shaped distribution over the width of the laser hardened track with a relatively small compressive RS plateau at about -100 MPa in the central section of the laser track. The transition zone is characterized by high compressive RS down to -600 MPa. Outside of the processed area the compressive RS are balanced by tensile RS. The decrease of the compressive RS in the center line of the track is due to a material distortion by volume expansion during martensite formation, i.e. the material flows evades towards the free surface and shows a deformation in this zone. The integral peak widths (see Fig. 3, right) are around 8.5° inside and 4.5° outside the process zone with a steep rise at the edges. This strong line broadening inside the laser treated region is a clear indication of increased dislocation density due to martensitic hardening.

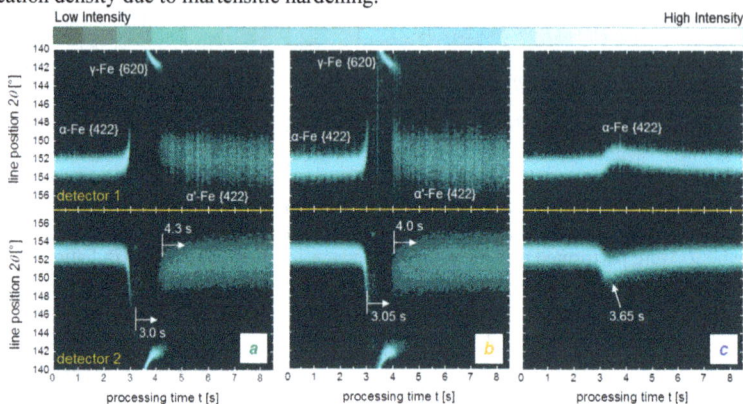

Fig 4: Normalized X-ray intensity plots recorded during the simultaneous in-situ diffraction analyses and laser hardening treatment. Shown are the results of detectors 1 and 2 at the three different measuring points y = 0 mm (left), 2 mm (middle) and 4 mm (right) of laser track (Fig. 1.). Reflections hkl are indexed in the upper detector 1 plots and the processing timings in the lower detector 2 plots.

In Fig. 4 normalized X-ray intensity plots recorded during the in-situ synchrotron X-ray diffraction studies that were taken simultaneously with the laser hardening treatment are presented for the three different measuring positions. At approx. 2.8 s a steep and high peak shift for the α-Fe {422} reflection occurs at the measuring positions *a* and *b* (left and middle), followed by the diffraction signal of supercooled {620} γ-Fe at around 3.7 s shifting to higher 2θ values due to increased supercooling up to 4.3 s (left) and 4.0 s (middle) when martensite formation starts. This is characterised by the appearance of the rather broad diffraction lines. The earlier martensite formation for position *b* (at 4.0 s, middle), compared to position *a* (4.3 s, left), can be attributed to the increased cooling rate due to higher degree of self-quenching towards the laser track edge. Outside the laser track at position *c* (Fig. 4, right) no phase transformation occurred. The maximum peak-shift and thus maximum temperature is reached at about 3.65 s. In contrast, at the given feed, the laser beam passes the 23 mm measuring point at 3.3 s. This time delay is a consequence of the time for heat conduction to occur. In Fig. 5 the thermal strains and deviatoric stresses are plotted versus the processing time for the three different measuring positions. The processing temperature T_{max} is plotted next to the thermal strain results. Additionally, results of the ex-situ RS measurements (see Fig. 3) are plotted next to the deviatoric stresses. Maximum thermal strain inside the process zone (positions *a* and *b*) for the heating of ferrite is reached directly before austenite formation starts. Due to the laser heat input that shows a Gaussian function for the

MECA SENS 2017   Materials Research Forum LLC
Materials Research Proceedings **4** (2018) 91-96   doi: http://dx.doi.org/10.21741/9781945291678-14

applied optics, the thermal strain is highest in the center of the laser track at 3.1 s when austenisation starts in the process zone. Outside the process zone (position c) the maximum thermal strain is delayed to 3.65 s and 3 to 5 times lower than in the process zone which is in good correlation with Fig. 4. With the onset of the martensite formation the thermal strain for position b decreases earlier and more rapidly (see Fig. 4) than for position a. This is a direct result of the higher local temperature gradient and thereby cooling rate at the edge of the process zone. The lowest cooling occurs at position c due to temperature balancing by heat conduction. Considering the stress-time plots for all three positions a - c, the initial stresses of about -40 MPa show a slight shift towards the tensile regime at e.g. about 2.8 s (position a), followed by a sharp decline towards high compressive stresses up to an approximate maximum of -600 MPa (position b) at about 3.05 s. The first increase can be explained by a global heating effect, which is almost identical for all three measuring positions. The later drop to high compressive stresses is a consequence of high thermal expansion and the restraint, which are different for the three measuring positions. The compressive stress decrease at measuring position c, after reaching a maximum of approximately

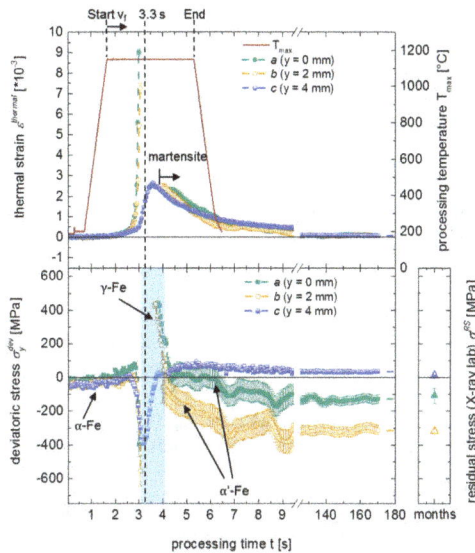

Fig 5: Thermal strain (top) and deviatoric (bottom) stress (transverse component) evolution during the laser line hardening process for the different measuring positions a, b and c. RS results according to Fig. 3 (bottom left). Light blue background indicates the austenite regime.

-400 MPa at about 3.15 s, is assumed to be related with austenite undercooling in the adjacent process zone. The high rate of volume contraction in the process zone during cooling combined with the requirement of material cohesion leads to small tensile stresses, which slightly decrease with the onset of martensite formation in the adjacent process zone to a near stress-free state (at about 4 s). The temporal stress development in the process zone (martensite formation) shows a decrease from about 100 MPa ($\approx$ 4.3 s) to approx. -100 MPa ($\approx$ 9 s) for measuring position a, which remains almost constant afterwards. Considering the thermal strain evolution in and outside the process zone, the decrease is a result of decreasing thermal strains. For measuring position b the stresses evolve to higher compressive residual stresses of about -300 MPa. This is much steeper than for position a due to the fact that volume expansion is locally more constrained. Since martensite formation starts at the edge and develops towards the center of the process zone and temperature is highest in the center, it is assumed that the material deforms plastically in the process zone. However, due to transient temperature distribution, the center is subjected to higher temperatures for a longer time interval that may cause local recovery, resulting in decreased integral widths compared to the edge of the process zone (see Fig. 3 left). The measured material deformation of the process zone supports this point. For all measuring positions at the end of the in-situ determined stresses, the values match well with the results of the RS lab measurements.

In comparison to previous work [6] no significant increase of the surface RS in the center of the laser track was measured. Main reason is the inhomogeneous (Gaussian) energy distribution of the laser beam optics and hence strong differences of local temperature, temperature gradients and cooling rate inside the process zone. The local and transient transformation behaviour may lead to higher degree of transformation plasticity in the track center and hence in lower compressive RS.

Conclusions

Spatial and time resolved synchrotron X-ray diffraction analysis during laser surface line hardening was successfully carried out. An improved experimental setup (two axes tiltable laser optics with in-line pyrometer) allowed for complete temperature controlled laser hardening processing with feed rates in line with industrial applications.

- Post processing stress measurements are in good agreement with ex-situ lab RS analyses.
- Thermal strain evolution is maximised in the center of the laser track decreasing towards the edge of the hardened zone due to Gaussian power distribution and corresponding heat input.
- Time of martensite formation depends on the position perpendicular to the track axis. Towards the laser track center the lower cooling gradient leads to later martensite formation.
- Local and transient martensite transformation in combination with local material deformation results in lower compressive RS in the track center that increases towards the edge of the process zone.
- Higher strain constraint at the edge of the process zone leads to increased plastic deformation and higher compressive RS compared to the center.

Acknowledgements

Financial Support by the German Research Foundation (DFG) in the projects GI376/10-1 and BE5341/1-1 is gratefully acknowledged.

References

[1] T. Miokovic, V. Schulze, O. Vöhringer and D. Löhe, Auswirkung zyklischer Temperaturänderungen beim Laserstrahlhärten auf den Randschichtzustand von vergütetem 42CrMo4, HTM 60 (2005) 142-149. https://doi.org/10.3139/105.100334

[2] K. Obergfell, V. Schulze and O. Vöhringer, Simulation of Phase Transformations and Temperature Profiles by Temperature Controlled Laser Hardening: Influence of Properties of Base Material, Surf. Eng. 19 (2003) 359-363. https://doi.org/10.1179/026708403225007572

[3] K. Müller and H.W. Bergmann, Suitability of materials for laser beam surface hardening, Z. Metallkunde, 90 (1999) 881-887.

[4] P. De la Cruz, M. Odén and T. Ericsson, Effect of laser hardening on the fatigue strength and fracture of a B–Mn steel, Int. J. Fatigue 20 (1998) 389-398. https://doi.org/10.1016/S0142-1123(98)00010-3

[5] E. Macherauch and P. Müller, Das $sin^2\psi$-Verfahren der röntgenographischen Spannungsermittlung, Z. angew. Phys. 13 (1961) 305-312.

[6] D. Kiefer, J. Gibmeier, F. Beckmann and F. Wilde, In-situ Monitoring of Laser Surface Line Hardening by Means of Synchrotron X-Ray Diffraction, Mat. Res. Proc. 2 (2016) 467-472.

[7] V. Kostov, J. Gibmeier, F. Wilde, P. Staron, R. Rössler and A. Wanner, Fast in situ phase and stress analysis during laser surface treatment: A synchrotron x-ray diffraction approach, Rev. Sci. Instrum. 83(11510) (2012) 1-11. https://doi.org/10.1063/1.4764532

MECA SENS 2017
Materials Research Proceedings 4 (2018) 97-102

Materials Research Forum LLC
doi: http://dx.doi.org/10.21741/9781945291678-15

Synchrotron XRD Evaluation of Residual Stresses Introduced by Laser Shock Peening for Steam Turbine Blade Applications

M. Newby[1,a*], A. Steuwer[2,b], D. Glaser[3,4,c], C. Polese[4,d],
D.G. Hattingh[5,e] and C. Gorny[6,f]

[1]Eskom Holdings SOC Ltd, Rosherville, Johannesburg, South Africa

[2]University of Malta, Msida MSD 2080, Malta

[3]CSIR National Laser Centre, Brummeria, Pretoria, South Africa

[4]University of the Witwatersrand, Johannesburg, South Africa

[5]Nelson Mandela University, Port Elizabeth, South Africa

[6]Laboratoire PIMM (ENSAM, CNRS, CNAM, Hesam Université), Paris, France

[a]mark.newby@eskom.co.za, [b]axel.steuwer@gmail.com, [c]dglaser@csir.co.za,
[d]claudia.polese@wits.ac.za, [e]danie.hattingh@mandela.ac.za, [f]cyril.gorny@ensam.eu

Keywords: Laser Shock Peening, Residual Stress, Synchrotron, Diffraction

Abstract. Steam turbines used in the power generation industry are subject to fatigue during normal operation which includes transient events such as start-ups and steady state operation. Surface treatment methods, such as shot peening (SP) and roller burnishing, to induce surface compressive residual stresses in critical areas and improve fatigue life are commonly used, but the depth of the induced residual stresses is limited by the process. Laser shock peening (LSP) is a more recent development that has been applied in the aerospace industry on titanium blades, but is not yet commonly used in the power generation industry. The current research is focused on optimizing LSP parameters for the application of the process on 12Cr steels used for turbine blades. Evaluation of the induced residual stress was done with both conventional laboratory X-ray diffraction (XRD) and synchrotron X-ray diffraction (SXRD) techniques.

Introduction

Thermal energy is the most commonly used source for electricity generation globally, which is often extracted by the use of large steam turbines. The corrosion resistant steel blades of a typical low pressure (LP) rotor operate in a wet steam environment, whilst rotating at speeds in the range of 3000 to 3600 rpm. The approximately 1m long LP blades therefore see high centrifugal loading, which presents challenges of stress corrosion cracking or corrosion fatigue [1]. The highly stressed fir-tree attachment root, as illustrated in Fig. 2, is conventionally shot peened (SP) in order to introduce beneficial compressive residual stresses to mitigate crack initiation. A catastrophic failure of one of these blades at a South African power station in 2003 resulted in over €100 million damage, and raised concerns to the effectiveness of the conventional SP treatment for the achievement of uniform coverage over the complex geometry of the fir tree section.

Laser shock peening (LSP) has been identified as an attractive technology for this application in order to potentially enhance the lifespan of the critical LP turbine blades. One of the benefits of a laser-based technology is the precise control of laser parameters, and hence the potential to introduce a beneficial compressive stress field to the desired level. Furthermore, the plasma

MECA SENS 2017   Materials Research Forum LLC
Materials Research Proceedings **4** (2018) 97-102   doi: http://dx.doi.org/10.21741/9781945291678-15

generated on the material surface produces a pressure normal to that surface which is potentially highly attractive to apply a uniform treatment to a complex 3D surface, such as a turbine blade fir-tree. The CSIR National Laser Centre (NLC) and Eskom are currently conducting research into the application of LSP for turbine blade refurbishment in close collaboration with SA universities.

The mechanism of the LSP process is depicted in the schematic in Fig.1. A high intensity laser pulse (duration in the order of nano-seconds) irradiates a metal target to rapidly ionize the surface into a plasma. When the plasma is confined with a medium transparent to the laser pulse, Giga Pascal (GPa) magnitude pressures are generated over a nano-second time frame which drives a shock wave through the metallic target. Plasticity is produced through the surface to a depth whereby the shockwaves no longer exceed the dynamic yield strength of the material. The material's elastic response to the plastic strains is the generation of a beneficial compressive residual stress to depths typically around 1 mm or greater. A sacrificial protective coating may be applied temporarily during LSP processing in order to prevent a direct laser-material interaction, which ensures a purely mechanical cold-working process. After each LSP application the tape is removed and replaced, thus these increments are generally referred to as tape layers. An alternative technique is to laser shock peen the surface without a protective coating (LSPwC), through a thin water containment film. An initial study conducted on 12Cr samples at the NLC yielded encouraging results for both LSP and LSPwC processes, however optical analysis of the spatial intensity profile across the laser spot showed significant variation from the desired top-hat profile, illustrated in the right hand images of Fig. 1. As a result a new batch of samples was processed for the results presented in this paper. These samples were treated at Laboratoire PIMM, in Paris.

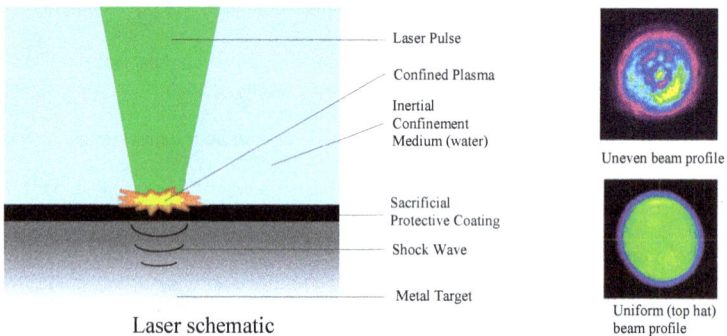

Laser Pulse

Confined Plasma

Inertial Confinement Medium (water)

Sacrificial Protective Coating

Shock Wave

Metal Target

Uneven beam profile

Uniform (top hat) beam profile

Laser schematic

Fig. 1: A schematic of the LSP process (left) and of two laser beam profiles (right).

Methodology

Sample Generation: Samples were extracted from an ex-service turbine blade by removing slices from the fir-tree attachment region as depicted in Fig. 2. A stress relieving cycle of 660°C for 20 minutes was performed on the coupons. The samples were wire EDM cut to dimensions of 20 x 20 x 15 mm^3 which were also surface ground. Electro-polishing was applied to remove the surface grinding effects, and hardness was checked to ensure that the stress relieving did not alter the mechanical properties. Laboratory XRD measurements were performed on each sample before and after LSP processing for repeatability verification of sample preparation and LSP processing.

MECA SENS 2017   Materials Research Forum LLC
Materials Research Proceedings 4 (2018) 97-102   doi: http://dx.doi.org/10.21741/9781945291678-15

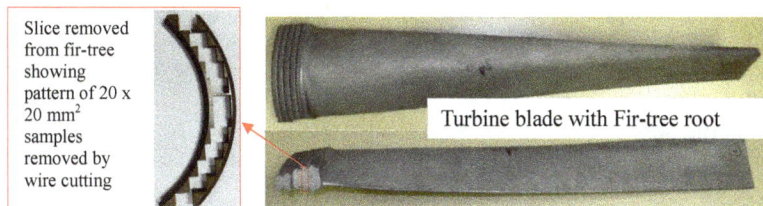

Fig. 2: Turbine blade configuration.

LSP and LSPwC Processing: The application of laser peening was performed at the PIMM Laboratory (ENSAM-CNRS-CNAM) using a Thales GAIA laser system operating at 532 nm with the sample immersed in a water tank. For the LSP work, a sacrificial thermo-protective overlay in the form of a black PVC tape was used (around 100 µm thick with a 30 µm adhesive). In order to ensure coating integrity, the spot-to-spot overlap was kept low at around 21.5 %. A spot size of 2.0 mm was selected, as this is approximately the largest diameter spot that would be practical to process within the turbine blade fir-tree as depicted in Fig. 2. A preliminary phase was first conducted in order to determine an appropriate power intensity whereby a constant spot size and overlap were used with varying power intensity from 1 to 8 GW/cm^2. Surface XRD measurements were made to identify 5 GW/cm^2 as a conservative power intensity operating below the saturation limit of the process due to dielectric breakdown before the target. The LSP processing was performed on the 20 x 20 mm^2 sample face with a 10 x 10 mm^2 LSP patch as depicted in Fig. 3. The LSPwC processing used two different spot sizes, 0.6 and 0.8 mm with coverage parameters of Np = 16.55 and 33.75, (Np = the number of pulses per mm^2).

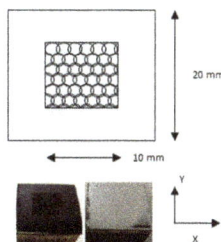

Fig. 3: LSP application pattern.

Laboratory X-ray diffraction: The surface residual stress analysis was performed using XRD measurements with a Proto iXRD instrument from Proto Manufacturing Inc., USA. A Cr-Kα X-ray source with a wavelength of 2.291 Å was used with a round 1.0, 0.5 or 0.2 mm aperture. Reflections from the (211) peak for the steel were used with a goniometer range of either ± 30° or ± 25° with a 3° oscillation, where the Bragg angle was 156.31°. The measurements were performed as per the sin2ψ technique with a minimum of 7 angles per strain measurement for the surface maps, for the depth profiles 11 angles per measurement were used to improve accuracy. Tri-axial measurements were performed at 0, 45, and 90 degrees in order to obtain the principal stresses. Stresses were calculated using the X-ray elastic constants -S1 = 1.15x10^{-6} MPa^{-1} and 1/2S2 = 5.247 x10^{-6} MPa^{-1}, determined through four-point bend tests.

Synchrotron X-ray diffraction: SXRD measurements were conducted at the ID15A beamline (experiment ME1440) at the ESRF facility in Grenoble, France. Energy dispersive measurements were performed with up to 300 keV energies which allowed for transmission through the 20 mm dimensions of the samples in order to provide strain measurements in the Y-direction as indicated in the schematic in Fig. 3. The beam dimensions were set to 50 µm by 100 µm, where the smaller dimension is in the depth (i.e. perpendicular to the LSP surface). A diffracting angle of around 3° results in gauge volume elongation to around 1.9 mm. A measurement time of two minutes was used for each position in the sample. The data was processed using GSAS Pawley

MECA SENS 2017   Materials Research Forum LLC
Materials Research Proceedings **4** (2018) 97-102   doi: http://dx.doi.org/10.21741/9781945291678-15

analysis in order to determine the lattice parameter accounting for multiple peaks for the bcc material. The strains were computed to stresses by assuming a bi-axial stress where an elastic modulus of 204 GPa and Poisson's ratio of 0.3 were used for computation. A small pillar of the material (2 x 2 mm^2 by 10 mm length) was used in order to determine the stress-free lattice parameter, d_{zero}.

Results and Discussion

Fig. 4 shows the XRD surface and SXRD through thickness residual stress results for three different power intensities with one layer of tape. Previous work [2, 3] has shown that the depth profiles become more consistent and deeper with the second tape layer and this is reflected in the data variation of up to 100 MPa peak to peak (pk-pk) shown in the plots. There is good correlation between the XRD and SXRD data at the surface of the samples, and a clear indication of the improved depth profile with increasing power intensity.

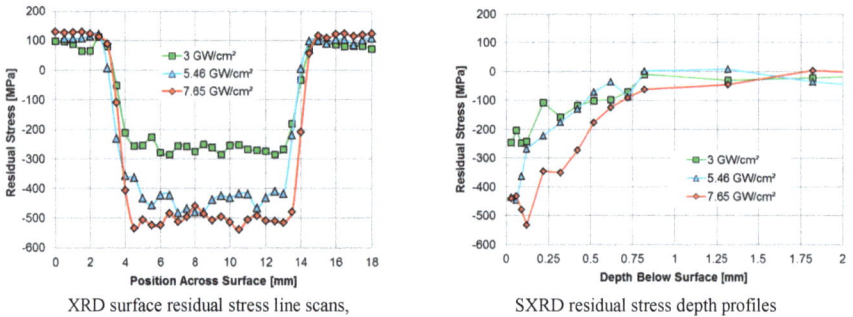

XRD surface residual stress line scans,
1 mm aperture, 0.5 mm spatial resolution

SXRD residual stress depth profiles

Fig. 4: Residual stresses comparisons as function of laser power intensity employed with the LSP processing using one tape layer.

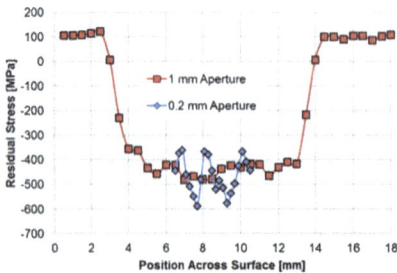

Fig. 5: Residual stress comparison from XRD measurements using two different aperture selections (1 tape layer).

Due to the observed variation of surface residual stress, further XRD measurements were performed with a 0.2 mm aperture as shown in Fig. 5. The averaging effect of the 1 mm aperture is significant for a 2 mm spot size. The 0.2 mm aperture highlights the variation in the surface stresses, which is periodic and correlates with the spot pattern. The pk-pk variation was approximately 220 MPa for a sample treated with 5 GW/cm^2 and one tape layer.

Fig. 6 shows the effect of two applications of the LSP process, illustrating that the second layer increases residual stress both at the surface and with depth. In addition the variation across the profile is improved significantly. In industrial applications a third layer is often used, but the previous work showed that for this material the third layer has limited benefit. The point at which the residual stress profile crosses zero improves from approximately 0.9 mm

MECA SENS 2017   Materials Research Forum LLC
Materials Research Proceedings **4** (2018) 97-102   doi: http://dx.doi.org/10.21741/9781945291678-15

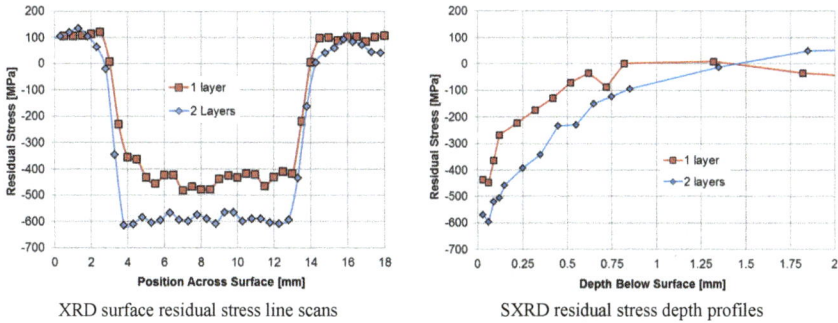

XRD surface residual stress line scans   SXRD residual stress depth profiles

Fig. 6: Effect of 1 or 2 tape layers on the induced residual stress values.

to 1.5 mm. Error bars are not shown for clarity of the plots, but typically the strain error was less than 70 µm/m.

Fig. 7 shows SXRD data for the LSPwC process with the surface stress value take from the XRD measurements. The samples had a rougher surface than LSP as shown in Fig. 9. The process results in a thin oxide layer on the surface and a recast layer below that. The effect of this on the SXRD data is that the d_{zero} in the first 100 µm is different from the bulk and this affects the data over the first 150 µm, due to the beam width. The plots in Fig. 7 are thus a combination of the two measurement processes and the focus is on the profile after 250 µm. The benefit of increased coverage in the LSPwC process is shown by the improved depth and amplitude of residual stress when Np increases from 16.85 to 33.75. Decreasing spot size has a benefit on the surface residual stress but the depth profile improves with the increase in spot size from 0.6 to 0.8 mm.

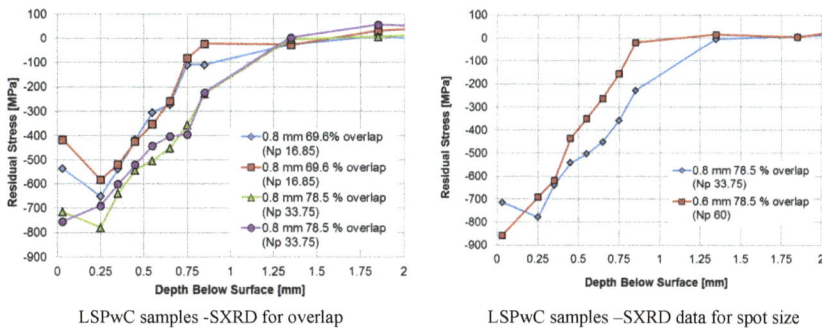

LSPwC samples -SXRD for overlap   LSPwC samples –SXRD data for spot size

Fig. 7: Coverage and spot size effect.

LSP and LSPwC both show a significant improvement over SP with the depth of residual stress at the zero crossing point increasing from 0.25 to 1.5 mm. LSPwC has a better depth profile than LSP, but the process is slower due to the smaller spot size and the higher overlap. In industrial applications the time required to treat a component will be one of the deciding parameters.

The benefit of improved surface roughness is clearly illustrated in Fig. 9, using the SP at 200 % coverage condition as a reference, the surface roughness Ra improves by a factor of 3.24 for LSPwC and 12.0 for LSP. This is particularly beneficial in environments where stress corrosion cracking can take place. The improved surface roughness will also aid in assembly of turbine blades into the location slots on the turbine rotor.

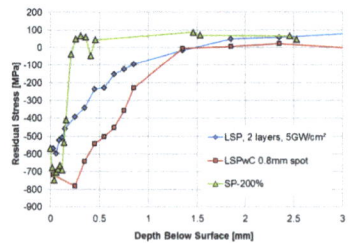

Fig. 8: Comparison of techniques

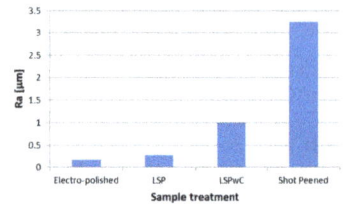

Fig. 9: Surface roughness.

Conclusions

The LSP and LSPwC treatment processes are both producing the desired results on 12Cr steel flat samples, with significantly improved residual stress profiles when compared to SP.

Residual stresses induced from LSP (2 mm spot, 21.4 % overlap, 5 GW/cm^2 and one tape layer) vary periodically by approximately ±220 MPa across the treated area.

The depth profile on a sample improves after two layers with treatment and the surface stress shows less variation. This is important when considering the time required for application on industrial components.

The residual stress transition from the treated to un-treated regions was very smooth and did not exhibit a sharp tensile peak.

Acknowledgements

The European Synchrotron Radiation Facility (ESRF) for beam time allocation, and assistance from Dr. Thomas Buslaps in experiment ME-1440.

References

[1] M. Newby, M. N. James and D. G. Hattingh, Finite element modelling of residual stresses in shot-peened steam turbine blades, Fatigue Fract. Eng. Mater. Struct. 37(7) (2014) 707–716. https://doi.org/10.1111/ffe.12165

[2] M. Newby, D. Glaser, and C. Polese, Laser Shock Peening Process Development for Turbine Blade Refurbishment Applications Using a Commercial 'Mid-Range' Energy Laser, in 6th International Conference on Laser Peening and Related Phenomena, 2016.

[3] K. R. Kuveya, C. Polese and M. Newby, Laser Peening versus Shot Peening Effects on Residual Stress and Surface Modification of X12CrNiMo12 Turbine Blade, in 6th International Conference on Laser Peening and Related Phenomena, 2016.

MECA SENS 2017   Materials Research Forum LLC
Materials Research Proceedings 4 (2018) 103-108   doi: http://dx.doi.org/10.21741/9781945291678-16

Neutron Diffraction Investigation of Residual Stresses in Nickel Based Austenitic Weldments on Creep Resistant Cr-Mo-V Material

P. Doubell[†,1,a*], M. Newby[1,b], D. Hattingh[2,c], A. Steuwer[2,3,d] and M.N. James[2,4,e]

[1]Eskom Holdings SOC Ltd, Lower Germiston Road, Rosherville, Johannesburg, South Africa

[2]Nelson Mandela Metropolitan University Port Elizabeth, South Africa

[3]University of Malta, Msida MSD 2080, Malta

[4]University of Plymouth, Drake Circus, Plymouth, UK

[a]philip.doubell@eskom.co.za, [b]mark.newby@eskom.co.za, [c]Danie.Hattingh@nmmu.ac.za,
[d]axel.steuwer@gmail.com, [e]m.james@plymouth.ac.uk

Keywords: Creep Resistant Alloy, Butt Weld, Stress Relief Crack, Residual Stress, Post-Weld Heat Treatment, Neutron Diffraction Strain Measurement

Abstract. Residual creep ductility of service-aged Cr-Mo-V creep resistant material is considerably lower than that of new material; this affects the long-term creep life performance of components manufactured from such alloys as the creep rate in aged alloy is considerably higher than for new materials. This study focused on the effects of residual stress and post-weld heat treatment (PWHT) on the remaining life of creep-exhausted material after repair welding using nickel-based consumables. Residual stresses attributed to the ferrite-to-austenite phase transformations involve a sudden volume change of the weld material. This can adversely affect aged material, e.g. a ½Cr-½Mo-¼V alloy, with low creep ductility and known notch sensitivity rendering this alloy prone to reheat cracking. Coupons prepared from creep damaged Cr-Mo-V pipes (323 mm outside diameter and 36 mm thick) were joined with the tungsten inert gas (TIG) and manual metal arc (MMA) welding processes simulating the original construction joints. Standard welding procedures were used with and without the addition of stress relief and temper post-weld heat treatment. Butt weld coupons were subsequently prepared, using a Ni-based consumable and a conventional ferritic consumable, for tri-axial stress measurements on the SALSA neutron diffraction beamline (ILL Grenoble), d_0 calibration used toothcomb specimens sectioned from the weld coupons. The industrial application of the experiments was sensitivity analysis of residual life prediction in FE modelling of plant system stresses in weld-repaired Cr-Mo-V creep resistant materials.

Introduction

The welding of high temperature and pressure components on power plants is considered a well-established and mature technology. It is used extensively during construction and maintenance activities. Maintenance activities, however, include welding and the effects of thermal aging of materials and components due to operation under load. These thermal effects can include phenomena such as a reduction in ductility, creep damage and high notch sensitivity that could render a component irreparable.

The chromium-molybdenum-vanadium (Cr-Mo-V) family of creep resistant alloys is known to be susceptible to weld and heat treatment related problems such as reheat cracking [1], also known as stress-relief cracking (SRC). This phenomenon is typified by cracks that occur in the coarse grained heat-affected zone, designated as a Type III crack in the categories proposed by Schüller [2], according to their location and geometry in a weld. Brett [3] subsequently added the

MECA SENS 2017   Materials Research Forum LLC
Materials Research Proceedings 4 (2018) 103-108   doi: http://dx.doi.org/10.21741/9781945291678-16

term Type IIIa crack which better describes the character of this intergranular SRC (See Fig. 1) found in the CGHAZ adjacent to the fusion line.

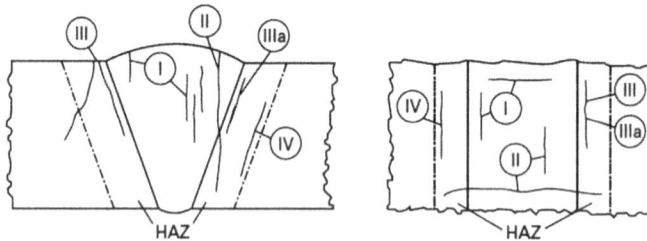

Fig. 1: Type IIIa reheat crack location [2].

SRC may occur when welds are exposed to high temperature during post-weld heat treatment or during operation at high temperature when creep ductility is not adequate to accommodate the elevation of strain due to residual stress [4].

Rationale for the application of Nickel-based weld consumables on ferritic alloys

Austenitic welding consumables are often proposed for application on ferritic base materials as an effective tool to alleviate the tendencies for reheat cracking and to extend creep life in refurbishment of components and support plant life-extension programmes. Limited information on the topic can be found in published literature on which to base a refurbishment strategy for nickel (Ni)-based alloys, in lieu of a matching ferritic filler material. Some information suggests that welding with Ni-based austenitic weld consumables could be beneficial in reducing weld-induced residual stresses. Nerger et al. reported a refurbishment project at the Hagenwerder Power Station in Germany to repair reheat cracking damage on butt welds joining Grade 14MoV6-3 pipes on the hot reheat pipe network [5]. Attempts to repair with ferritic weld consumables followed by PWHT resulted in extensive new cracking occurring in the CGHAZ. The solution adopted was to perform the weld repairs with a temper-bead technique using Ni-based weld consumables without subsequent PWHT. Depositing Ni-base electrodes reduces the off-loading effects arising from matched strength electrodes. Such electrodes create a narrower heat-affected zone (HAZ) than ferritic electrodes of the same size. Another motivation for utilising Ni-based consumables is to achieve a relatively lower stress weld as no phase transformation occurs in the weld metal solid phase due to stability of the austenitic phase from low temperatures up to the melting point.

No phase transformation stresses occur, giving lower thermal stresses during the weld cycle when compared with ferritic consumables where ferrite-to-austenite transformation involves a sudden volume change of the weld material. This can have an adverse effect on aged material with low creep ductility. When the notch sensitivity of an alloy such as Grade 14MoV6-3 is added into the equation, this group of alloys becomes prone to the reheat cracking phenomenon (Fig. 2).

The Ni-base weld metal generally has better toughness than a ferritic weld metal, while the higher solubility of hydrogen in austenitic weld metal is thought to alleviate the risk of hydrogen-assisted cracking.

A disadvantage of using nickel-base filler metal is that the dissimilar metal combination considerably hinders any NDE by ultrasonic-testing methods, due to the difference in ultrasonic wave-propagation characteristics, while magnetic particle testing cannot be used as a surface crack-detection method. Notwithstanding these issues, the potential benefits for using nickel as a part of a repair strategy purportedly outweigh the disadvantages when applied correctly and supported by sufficient engineering assessment.

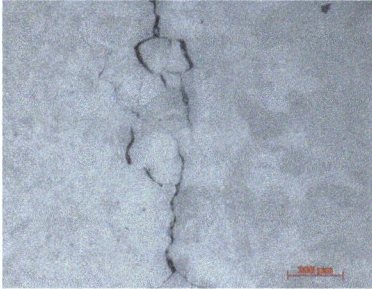

| Fig. 2: Reheat cracking damage in weld heat-affected zone. | Fig. 3: Surface replication for determining microstructure integrity. |

Plant life extension calculations and repairs rely on in-situ evaluation techniques such as surface replication (Fig. 3) to detect the possible incidence of creep damage, hydrogen cracks and SRC, an approach that supports efforts to determine component integrity down to microstructural level. This study investigated some of these claims regarding the virtues of Ni-base weld consumables, in particular the claims of a lower stressed weld condition compared with using ferritic weld consumables that match the base alloy.

Welding of test pieces and mechanical tests
Due to the relatively wide potential applications of so-called reduced stress welding techniques, it was decided to focus on one particular application often encountered during maintenance activities i.e. spool-piece replacement on the main steam and re-heater pipework, typically manufactured from creep resistant alloys to BS EN 10216-2 Grade 14MoV6-3 which is known for high levels of notch sensitivity. Table 1 lists the nominal chemical constituents for this alloy.

Table 1: Chemical composition for Grade 14MoV6-3 creep resistant alloy

C	Cr	Mo	V	Si	Mn	P	S
0.1-0.18	0.3-0.6	0.5-0.7	0.22-0.32	0.1-0.35	0.4-0.7	0.035 max	0.035 max

Spool-pieces were prepared from retired creep damaged Cr-Mo-V pipes, nominally 323 mm outside diameter and 36 mm thick, joined with the TIG and MMA welding processes. For this study conventional ferritic 2¼%Cr-1%Mo weld consumables were used for one coupon while the second coupon used a 70%Ni-19%Cr-5%Mn-2%Nb-1.5%Mo-3%Fe type austenitic weld consumable. Standard welding procedures were followed both with, and without, additional stress relief and temper PWHT for comparison.

Residual stress measurement
The aim of the experiment was to make neutron diffraction residual stress measurements in four test coupons made with either matching ferritic or Ni-based weld consumables to validate the heat-treatment processes (see Table 2). The measurements were performed at the Institut Laue-Langevin (ILL) on the Strain Analyser for Large Scale Engineering Applications (SALSA)

MECA SENS 2017   Materials Research Forum LLC
Materials Research Proceedings **4** (2018) 103-108   doi: http://dx.doi.org/10.21741/9781945291678-16

beamline in the five days allocated for experiment 1-02-31. Fig. 4 shows the general layout of the test cubicle with a coupon mounted on the hexapod table. Neutron diffraction is the only technique suitable for measuring residual strains in the thick steel samples used in this experiment. A positional matrix comprising 31 points was originally planned for each sample; however, when the experiment started it was evident that weld texture led to extremely long measurement times and the matrix had to be reduced to three lines for the as-welded samples (5, 16 and 35 mm) and one line (16 mm) for the heat-treated samples as shown in Fig. 5.

Fig. 4: SALSA test cubicle layout.

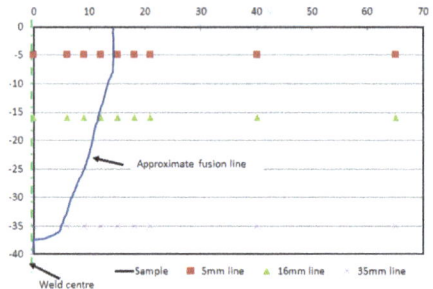

Fig. 5: Matrix of measurement points.

Through this experimental work the residual strains in four coupons in three directions (hoop, axial and radial) were determined.

Table 2: Test coupon welded and-heat treated conditions

Sample Number	Description
1A	Butt weld with Ni-based austenitic consumables without PWHT (As-welded)
1B	Butt weld with Ni-based austenitic consumables with PWHT
2A	Butt weld with ferritic consumables without PWHT (As-welded)
2B	Butt weld with ferritic consumables with PWHT
d_{zero} cubes	2 x 2 x 2 mm^3 cubes across weld zone for unstrained lattice parameter determination

Residual stresses were inferred from standard strain-to-stress formulae by assuming that the measured strain values were in the three principal directions and using an elastic modulus (E) of 212.7 GPa and a Poisson's ratio (v) of 0.28. The reference lattice parameter d_{zero} was measured on cubes 2 x 2 x 2 mm^3 extracted from sectioned weld coupons. The data obtained provide critical information relevant to the structural integrity of weld repaired Cr-Mo-V creep resistant materials and can be compared with FE modelling predictions. The SALSA instrument offers the ability to efficiently characterise stresses in weldments of 36 mm thickness at high spatial resolution. Radial collimators were used that provided a gauge volume of 0.6 x 0.6 x 2 mm^3 in conjunction with a neutron beam wavelength of $\lambda = 1.644$ Å.

Results and discussion
Notable results of the measurements are shown in the following figures:

MECA SENS 2017   Materials Research Forum LLC
Materials Research Proceedings **4** (2018) 103-108   doi: http://dx.doi.org/10.21741/9781945291678-16

- Fig. 6, comparison of as-welded and heat-treated data on the 16 mm line for samples 1A and 1B
- Fig. 7, comparison of as-welded and heat-treated data on the 16 mm line for samples 2A and 2B
- Figs. 8 and 9, surface contour plots extrapolated from the line scans on sample 2A.

Error bars have been omitted from Figs. 6 and 7 for clarity, but were in the order of ±50 μm/m in the ferritic material and ±110 μm/m in the austenitic material. The errors improved in the heat treated samples

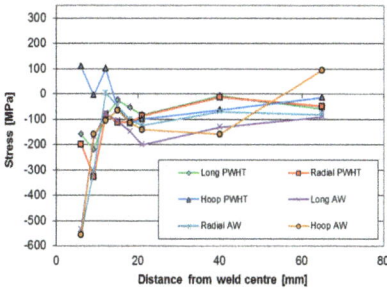

Fig. 6: Stress profiles 16 mm into sample, comparing 1A [As-welded] with 1B [Heat treated].

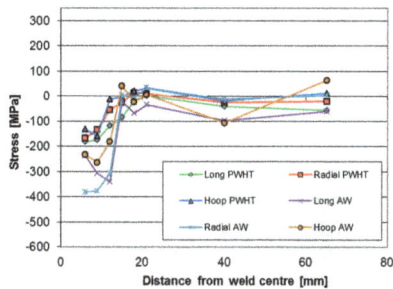

Fig. 7: Stress profiles 16 mm into sample, comparing 2A [As-welded] with 2B [heat treated].

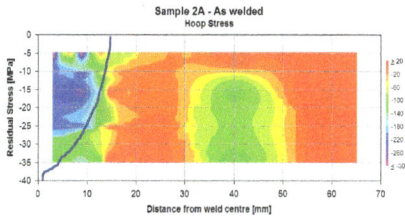

Fig. 8: Sample 2A - Extrapolated contour plot of hoop stress.

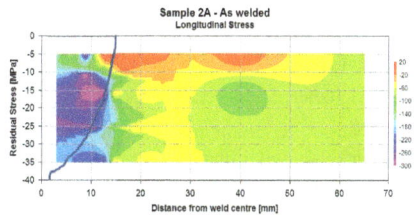

Fig. 9: Sample 2A - Extrapolated contour plot of longitudinal stress.

The experimental data for the as-welded condition of the austenitic weld metal on a ferritic base revealed a peak compressive stress exceeding the yield point of the Grade 14MoV6-3 base material in the vicinity of the HAZ. PWHT dramatically reduces this to a more gentle compressive residual stress gradient across the HAZ, the hoop stress appears to invert to a tensile value. The stress measurements on the ferritic weld consumable sample in the as-welded condition revealed a reduced residual compressive stress field compared to the sample welded with austenitic weld consumables. Predictably, PWHT reduced this to a lower compressive stress value. A comparison of the ferritic and austenitic consumable welds in the as-welded condition shows a significant difference in peak compressive levels. PWHT reduces this difference and it

MECA SENS 2017   Materials Research Forum LLC
Materials Research Proceedings **4** (2018) 103-108   doi: http://dx.doi.org/10.21741/9781945291678-16

can be assumed that the peak compressive stress fields for the two consumable types are the same for all practical purposes.

The results of the experiment clearly demonstrated the beneficial influence of PWHT by reducing the residual stress state of the completed welds for both ferritic and austenitic weld metal. From a weld residual stress point of view no conclusive evidence were found after analysis of the experimental data that can be construed as sufficient proof that austenitic weld consumables deposited on ferritic base material has significant metallurgical and mechanical benefits over ferritic weld consumables. However, the experimental methods applied here will however not be able to demonstrate the possible plant operational benefits that can be achieved during the weld thermal cycle when applying austenitic weld consumables to ferritic base materials.

Conclusions

- The benefit of heat treatment is clearly illustrated by the neutron diffraction tests
- The nickel-based consumable has higher as-welded stresses compared to the Fe filler
- No conclusive evidence could be found to prove that austenitic weld consumables deposited on ferritic base material has significant metallurgical and mechanical benefits over ferritic weld consumables
- The results of this study will support Eskom's life extension programme and the refurbishment of creep exhausted components

References

[1] C. Lundin and K. Khan, Fundamental Studies of the Metallurgical Causes and Mitigation of Reheat Cracking in 1¼Cr-½Mo and 2¼Cr-1Mo Steels. WRC Bulletin 409, February (1996)

[2] H. Schüller, L. Hagn and A. Woitscheck, Risse im Schweissnahtbereich von Formstücken aus Heissdampfleitungen, Werkstoffuntersuchungen; Der Maschieneschaden 47(1) (1974) 1-13.

[3] S. Brett, Type IIIa cracking in ½CrMoV steam pipework systems, Sci. Technol. Weld. Joining 9(1) (2004) 41-45.

[4] C. Meitzner, Stress relief cracking in Steel Weldments. WRC Bulletin (1975)

[5] D. Nerger, R. Blume and H. Schinkel, Weld reconditioning of a 14MoV 6 3 Hot Reheat Line with Nickel-Based Filler Metals without Subsequent Heat Treatment. VGB Kraftwerktechnik 74(10) (1994) 751-756.

Surface Modification and Coating

MECA SENS 2017
Materials Research Proceedings **4** (2018) 111-116

Materials Research Forum LLC
doi: http://dx.doi.org/10.21741/9781945291678-17

Neutron Through-Thickness Stress Measurements in Two-Phase Coatings with High Spatial Resolution

V. Luzin[1,a*] and D. Fraser[2,b]

[1]Australian Nuclear Science and Technology Organisation, Lucas Heights, NSW, 2232 Australia

[2]Commonwealth Scientific and Industrial Research Organisation, Clayton, Melbourne, VIC, 3168, Australia

[a]vladimir.luzin@ansto.gov.au, [b]Darren.Fraser@csiro.au

Keywords: Residual Stress, Coatings, Cold Spray

Abstract. Neutron diffraction residual stress profiling of sprayed coatings with high spatial resolution is a difficult task. Normally, only for single-phase materials 0.1 - 0.2 mm resolution can be achieved. Stress measurements in two-phase or multi-phase coatings are an even more formidable experimental task due to the necessity of measuring all phases with lower than 100 % volume fractions and the necessity to resolve the d_0 problem in a more complex way than the for single-phase coating systems.

The results of through-thickness residual stress profiling neutron diffraction experiments are reported on a selected two-phase, metal-metal composite coating, deposited by the cold spray technique. With both phase strains measured and complemented by additional information provided by other characterisations, the full stress state was reconstructed. Macro- and micro-stresses were separated allowing interpretation of the experimental data in terms of macro- and micro-mechanics. It also allowed us to make conclusions about the thermal mechanisms of macro- and micro-stress formation, as well as connection of these mechanisms to spraying parameters.

Introduction

A range of spray technologies are in use now for the purpose of surface enhancement and are frequently employed to produce coatings on the surface of numerous engineering components. Coatings of many different materials are produced by various spray techniques for numerous applications, such as wear resistance, corrosion protection, insulation, etc., and usually the spraying conditions and parameters are optimized for the purpose. One of the parameters to consider in such optimisation is the residual stress which is formed due to high temperatures or/and high kinetic energies associated with the spraying process, as well as mismatches in the substrate and coating material properties. Since the residual stress, which can negatively influence the coating's mechanical integrity or functional performance, stress control and mitigation are usually an integral part of the technology.

Neutron stress measurement in thick and thin coatings has proven to be a useful method since it is non-destructive, it can provide the required high resolution (down to 0.2 mm), it does not require special sample preparation (e.g. cutting and polishing, as for X-rays), measurement can be done in a reasonable time (minutes per datum) and with high accuracy (uncertainty can be better than 5 MPa). However, some cases remain challenging, e.g. 0.1 mm thick coatings, and another challenge is stress measurement in two-phase coatings, especially when the volume fraction of one of the phases is small.

In this work we report an experimental study of the residual stress analysis in the two-phase (metal-metal) coating-on-substrate system produced by the cold-spray technique.

MECA SENS 2017   Materials Research Forum LLC
Materials Research Proceedings 4 (2018) 111-116   doi: http://dx.doi.org/10.21741/9781945291678-17

Specifics of the stress measurements in two-phase coatings

Details of stress measurements of two-phase systems were outlined in [1], only a few key points are highlighted here. Macrostress profiling can be achieved through measurements of both phases and averaging phase stresses with the corresponding volume fractions, which can be found in a separate (e.g. neutron powder diffraction) experiment. As in a single-phase coating experiment, measurements of the strains in the in-plane and normal directions are required. There is no need for knowing exact d0 values if only the macrostress is of interest. Regarding the microstress, if no stress-free, d_0-powders are available, only the deviatoric part of the micro-stress tensor can be found. If, however, the d_0-powders (or at least one powder) are available, the hydrostatic part can be derived from the analysis of strain data.

Samples

The powder for spraying has chemical composition typical for the lean duplex stainless steels, though with somewhat high Mn content, so that the sprayed material has both fcc (γ-Fe) and bcc (α-Fe) iron. The chemical and phase composition was developed for a special (undisclosed) application requiring high temperature oxidation resistance and strength. Size distribution of the multi-disperse spherical powder particles, with diameter being within the range 5 to 50 μm (Fig. 1), was quantified resulting in the average particle size of approximately 23 μm. The details of the chemical compositions are given in Table 1.

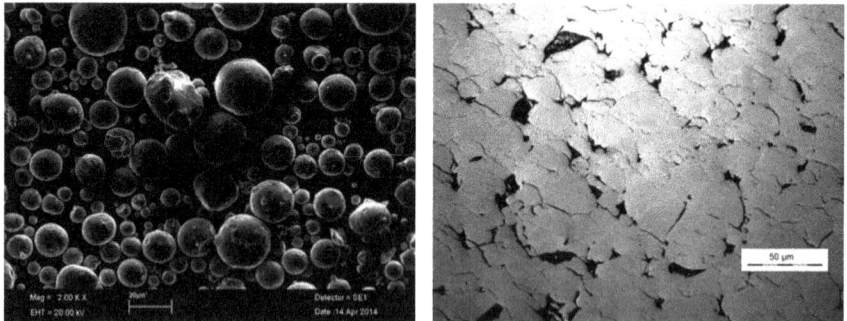

Fig. 1: Morphology and sizes of particles of the alloy powder (left) and microstructure of the sprayed material, sample #1 (right). Dark areas of the microstructure correspond to porosity.

Table 1. Chemical composition of the initial powders, wt.%.

	C	Mn	Cr	Mo	Ni	Co	Si
duplex stainless steel	<0.03	8.7	21.7	<0.1	4.4	4.2	0.37

To produce samples for the study, the powder was cold sprayed using the CSIRO's high pressure system, CGT Kinetiks 4000. While the gas (nozzle) temperature and pressure were fixed at 800°C and 35 bars respectively, some spray parameters were varied accordingly to Table 2. Relatively high temperature for the cold-spray gas technique was required to provide deposition conditions for the powder particles, as the particle size was relatively large for the cold spray. The combinations of spraying parameters were chosen in such a way that under assumption of constant deposition efficiency the expected thickness of coatings should be 3.0 mm. It was evident that for different regimes (sets of spray parameters, Table 2) the efficiency was not the same resulting in coatings of different thicknesses. All coatings were sprayed on

MECA SENS 2017   Materials Research Forum LLC
Materials Research Proceedings **4** (2018) 111-116   doi: http://dx.doi.org/10.21741/9781945291678-17

similar sized substrate patches, i.e. squares 40×40 mm^2 and 6.2 mm in thickness, machined from austenitic stainless steel 316.

Table 2. Sample spraying parameters.

Sample ID	powder	Feed rate, [kg/hour]	Traverse speed, [mm/s]	Passes	Thickness, [mm]
#1	duplex steel	1	300	60	3.1
#2	duplex steel	2	300	30	2.7
#3	duplex steel	1	100	20	1.5

Elastic properties and density evaluation
For the Young's modulus measurements, rectangular specimens were extracted from the bulk of the coating with approximate dimensions $38 \times 5 \times 2$ mm^3. The Young's modulus was determined using the Impulse Excitation Technique (IET) according to the ASTM standard E1876 through acoustic measurements of the normal frequency. For the given sample dimensions the accuracy of this method was generally better than 1 %. The same samples with highly accurate dimensions were used to evaluate their density through volume calculation and sample weighing. The results of the evaluations are reported in Table 3.

Phase composition measurements
The full neutron diffraction patterns were measured using the same bar samples for the evaluation of phase composition. Being a volumetric technique, neutron diffraction provided highly reliable bulk averaged volume fraction for bcc vs fcc. The diffraction patterns were measured over an angular range of 10 - 160° at a wavelength of 1.622 Å using the high resolution powder diffractometer ECHIDNA at the ANSTO OPAL research reactor [2]. The volume fractions of the phases were determined using the GSAS Rietveld refinement software [3] with the EXPGUI interface [4].The results of the determinations are reported in Table 3.

Neutron residual stress measurements
Neutron diffraction residual stresses measurements have been carried out using the stress diffractometer KOWARI at the ANSTO OPAL research reactor [5]. For through-thickness stress measurements a gauge volume with dimensions 0.5 x 0.5 x 20 mm^3 was used. The gauge volume was small enough to provide the necessary through-thickness resolution while producing a count rate sufficiently high for determining strain measurement in each phase with statistical uncertainty better than 5×10^{-5} within a reasonable measurement time. The exposure times were ~1 minute per position for the measurements in the substrate material and 5 minutes per point for the coating material due to significant effect of peak broadening. All measurements were done with a neutron beam wavelength of 1.54 Å providing approximately 90°-geometry for the two reflections that investigated, i.e. γ-Fe(311) and α-Fe(211) with the diffraction angles being 90° and 82° respectively.

The strain measurements were carried in many through-thickness positions covering the entire sample thickness. The 0.5 mm spacing between points was chosen to be proportional to the overall thickness of ~9 mm and gauge volume size of 0.5 mm. Notwithstanding the equi-biaxial stress state most likely to be expected, two in-plane directions and one normal direction were measured in order to reconstruct two in-plane stress principal components under the assumption of plane (macro-) stress condition.

A "substrate only" sample was measured with the same measurement protocol confirming the absence of any pre-existing stress distribution in the substrate material (e.g., from the production stage).

The experimentally determined phase d-spacing data were treated according to the stress reconstruction procedure [1] resulting in the macrostress through-thickness profiles (Fig. 3).

The macrostress, determined through the rule-of-mixture, $\sigma^M = f_\alpha \sigma_\alpha^t + f_\gamma \sigma_\gamma^t$, where σ_α^t and σ_γ^t are total phase stresses, was further treated within an empirical stress formation model, the progressive layer deposition model by Tsui & Clyne [6]. The thermal mismatch term, $\Delta\varepsilon = \Delta\alpha\Delta T$, due to the difference $\Delta\alpha$ between the coefficients of thermal expansion (CTE) of the substrate material (SS316) and coating material (α-Fe/γ-Fe composite), was found (as a best fit) to be the same for all three samples, $\Delta\varepsilon = -156 \pm 5$ µstrain. This corresponds to a temperature drop after spraying $\Delta T_1 \sim 380$ - $400°C$, making some assumption about the CTE of the coating phases, α-Fe and γ-Fe. The deposition stress was thus sample dependent and reported in Table 3.

The deviatoric part of the microstress tensor was also derived and shown in Fig. 4. Since it was impossible to obtain d_0-powders for any of the coating constituents, the hydrostatic part of the microstress was not determined experimentally. Instead, it was evaluated in a model approach, e.g. within Hashin-Shtrikman bounds theory [7], knowing that the hydrostatic part can only be generated thermally, taking into account the experimentally known phase compositions, assuming typical CTEs of α-Fe and γ-Fe and temperature drop $\Delta T_2 \sim 700°C$. The results of the total stress evaluation are shown in Fig. 4.

Table 3. Sample characterization: elastic properties, phase composition and stress analysis.

Sample ID	Young's modulus, [GPa]	Density, [g/mm³]	Volume fractions, [aus/ferr]	Deposition stress, [MPa]
#1	129.6 ± 0.5	7.259 ± 0.034	0.44/0.56	$+11 \pm 20$
#2	125.4 ± 0.7	7.237 ± 0.045	0.46/0.54	$+53 \pm 20$
#3	115 ± 2.5	6.858 ± 0.15	0.53/0.47	$+99 \pm 20$

Discussion

Macrostress. In all three samples there were two contributions to the macrostress, i.e. the thermal mismatch and the deposition stress. While thermal mismatch generated compressive stress in the coating, the deposition stress is tensile. The fact that the resultant stress is still compressive illustrates the dominant role of the thermal mismatch mechanism. Contrary to expectation of a compressive deposition stress, typical for the cold-spray materials [8], it was tensile, which is more typical for thermal spray techniques like plasma or HVOF (High Velocity Oxi-Fuel). Since the sign of the deposition stress is determined by a balance between the peening mechanism (compressive stress) and the quenching mechanism (tensile stress), our results suggest a much more significant role of quenching than peening, thus making the given regime of the cold-spray system being more reminiscent to HVOF.

Among the samples, the deposition stress parameter demonstrated a clear trend being more tensile for sample #3 and less tensile for sample #1. Assuming the above discussed concept of the peening/quenching balance, sample #1 exhibits the most amount of peening that is also reflected in a higher value of the Young's modulus, density and efficiency (accumulated thickness) due to better compaction of splats in the coating. The opposite is valid for sample #3.

Anti-correlation of the deposition stress and number of passes makes a link to the spray parameters: for sample #3 with only 20 passes, the localized heat-input must be greater than for the faster moving heat source for sample #1 with 60 passes, which allows more efficient heat dissipation and, therefore, reducing the role of the thermal/quenching effects.

Microstress. The hydrostatic microstress is clearly thermally generated, due to the difference between CTE of α-Fe and γ-Fe, and it is tensile in γ-Fe and compressive in α-Fe. Since the phase

MECA SENS 2017   Materials Research Forum LLC
Materials Research Proceedings **4** (2018) 111-116   doi: http://dx.doi.org/10.21741/9781945291678-17

composition for the three samples are close, the microstress is approximately the same, approximately 400 MPa in γ-Fe and -400 MPa in α-Fe.

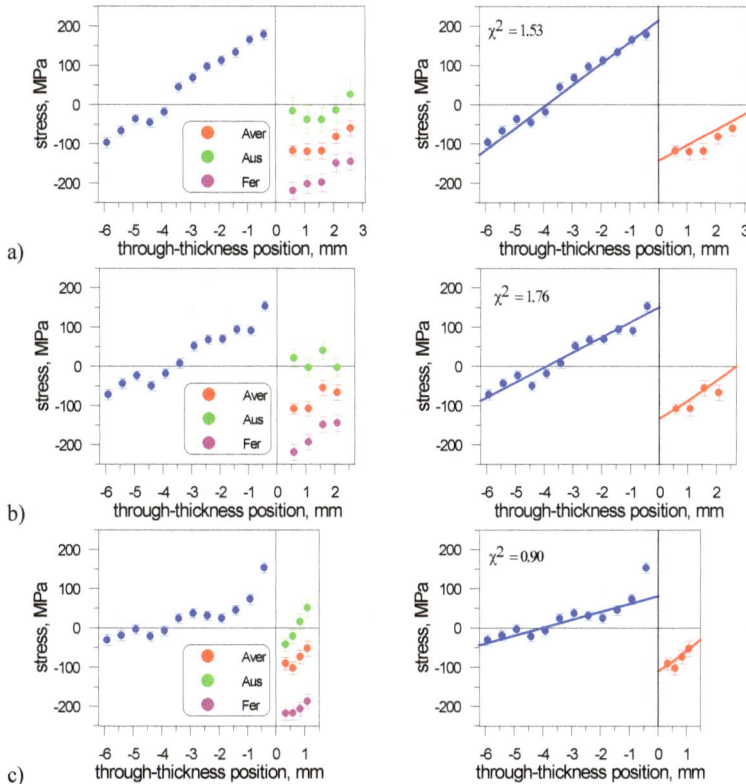

Fig. 3: Experimentally measured phase stresses (left column) and model fit of the macrostress profiles (right column) for three samples #1, #2 and #3, correspondingly a), b) and c). Aus represents the austenite phase and Fer the ferrite phase respectively.

The results show clear stress relaxation for the normal component – a reason for appearance of the deviatoric component. Two micro-mechanical mechanisms can be responsible, (i) oriented microcracking with the crack plane being parallel to the surface and (ii) plastic mismatch which is manifested in a differential amount of plastic deformation in the two phases, more plastic fcc γ-Fe vs less plastically hard bcc α-Fe. Considering the real microstructure (Fig. 1) exhibiting approximately round splats (assuming a small amount of the overall plastic deformation, but most likely large in localized areas) and visible porosity, the first mechanism (i) is most plausible, though with some presence of mechanism (ii). This also can be supported by a correlation between the amount of stress relaxation in the normal direction (largest in #3) and deterioration of the Young's modulus (largest in #3), which is also sensitive to microcracking.

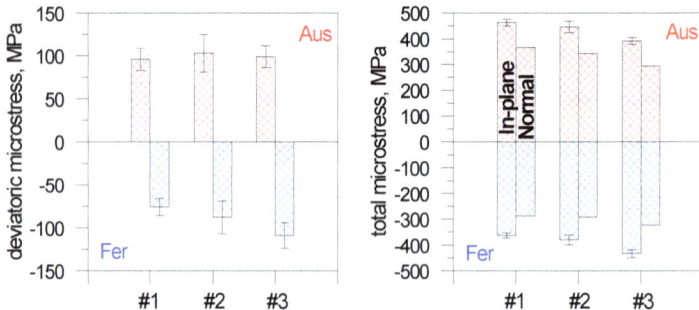

Fig. 4. Deviatoric part of the microstress (left, as measured) and the total microstress (right) obtained by adding the hydrostatic part (calculated). Aus represents the austenite phase and Fer the ferrite phase respectively.

Conclusions

Three different composite (γ-Fe/α-Fe) coating samples were sprayed by a cold-spray system in a high temperature regime, thus being closer to the thermal spray conditions but without melting. Stresses were measured in both phases with neutron diffraction with high spatial resolution of 0.5 mm. It was sufficient to resolve stress profiles in 1.5 - 3.1 mm thick coatings and make quantitative analysis possible. Experimentally determined macrostresses and microstresses were matched with modelling results to reveal the macro- and micro-mechanical mechanisms of stress formation, dominated in both cases by thermal nature.

References

[1] V. Luzin, J. Matejicek and G.-H. T., Through-thickness Residual Stress Measurement by Neutron Diffraction in Cu+W Plasma Spray Coatings, Mater. Sci. Forum 652 (2010) 50-56. https://doi.org/10.4028/www.scientific.net/MSF.652.50

[2] K.-D. Liss, B. Hunter, M. Hagen, T. Noakes and S. Kennedy, Echidna – the new high-resolution powder diffractometer being built at OPAL, Physica B: Condensed Matter 385–386 Part 2 (2006) 1010-1012. https://doi.org/10.1016/j.physb.2006.05.322

[3] A.C. Larson and R.B. Von Dreele, in: Report LAUR 86-748, Los Alamos National Laboratory, 2004.

[4] B. Toby, EXPGUI, a graphical user interface for GSAS, J. Appl. Crystallogr. 34 (2001) 210-213. https://doi.org/10.1107/S0021889801002242

[5] O. Kirstein, V. Luzin and U. Garbe, The Strain-Scanning Diffractometer Kowari, Neutron News, 20 (2009) 34-36. https://doi.org/10.1080/10448630903241175

[6] Y.C. Tsui and T.W. Clyne, An analytical model for predicting residual stresses in progressively deposited coatings Planar geometry, Thin Solid Films 306 (1997) 23-33. https://doi.org/10.1016/S0040-6090(97)00199-5

[7] Z. Hashin, The Elastic Moduli of Heterogeneous Materials, J. Appl. Mech. 29 (1962) 143-150. https://doi.org/10.1115/1.3636446

[8] V. Luzin, K. Spencer, M. Zhang, N. Matthews, J. Davis and M. Saleh, Residual Stresses in Cold Spray Coatings, in: P. Cavaliere (Ed.) Cold-Spray Coatings – Recent Trends and Future Perspectives, Springer, 2018. https://doi.org/10.1007/978-3-319-67183-3_16

MECA SENS 2017
Materials Research Proceedings 4 (2018) 117-122

Materials Research Forum LLC
doi: http://dx.doi.org/10.21741/9781945291678-18

Measurement of Residual Stresses in Different Thicknesses of Laser Shock Peened Aluminium Alloy Samples

S.N van Staden[1,2,a*], C. Polese[1,2,b], D. Glaser[3,c], J.-P. Nobre[1,2,d], A.M. Venter[2,4,e], D. Marais[4,f], J. Okasinski[5,g] and J.-S. Park[5,h]

[1]School of Mechanical, Industrial and Aeronautical Engineering, University of the Witwatersrand, 1 Jan Smuts Avenue, Johannesburg, 2000, South Africa

[2]DST-NRF Centre of Excellence in Strong Materials, University of the Witwatersrand, 1 Jan Smuts Avenue, Johannesburg, 2000, South Africa

[3]CSIR National Laser Centre, Meiring Naudé Road, Pretoria, 0184, South Africa

[4]Research and Development Division, Necsa SOC Limited, R104 Pelindaba, Pretoria, 0240, South Africa

[5]Advanced Photon Source, Argonne National Laboratory, 9700 S. Cass Avenue, Lemont, Illinois, 60439, USA

[a]sean.vanstaden2@students.wits.ac.za, [b]claudia.polese@wits.ac.za, [c]dglaser@csir.co.za, [d]joaopaulo.nobre@wits.ac.za, [e]andrew.venter@necsa.co.za, [f]deon.marais@necsa.co.za, [g]okasinski@aps.anl.gov, [h]parkjs@aps.anl.gov

Keywords: Laser Shock Peening, Aluminium, Residual Stress, Incremental Hole Drilling, X-ray Diffraction, Synchrotron, Energy-Dispersive, Neutron Diffraction

Abstract. This study focused on depth-resolved residual stress results determined with a number of complementary techniques on Laser Shock Peening (LSP) treated aluminium alloy 7075-T651 samples with different thicknesses (6 mm and 1.6 mm). Samples were prepared from a single commercially produced rolled plate that was then treated with LSP. Residual stresses were measured using Laboratory X-Ray Diffraction (LXRD), Incremental Hole Drilling (IHD), Neutron Diffraction (ND) and Synchrotron XRD (SXRD). The LSP treatment resulted in the establishment of compressive residual stresses that varied rapidly in the near surface region. The compressive stresses extended up to 1.5 mm in depth in the 6 mm thick sample. Some surface stress relaxation was observed in the first 25 μm, but substantially large stresses existed at 50 μm. This investigation strongly motivated why residual stress profiles should be obtained using a variety of techniques.

Introduction

Laser Shock Peening (LSP) is a surface treatment technique that involves the ablation of a metal sample by pulsed, high intensity, laser irradiation. The sample surface is covered by some medium transparent to the laser (typically water). A plasma is formed at the surface due to the rapid heating by the laser. This event is confined by the transparent medium so, as the plasma expands, it generates extremely high pressures which are transferred to the sample through shock waves. These plastically deform the sample and are expected to establish a beneficial compressive residual stress near the surface [1].

Such a compressive residual stress field can result in slowed fatigue crack propagation and increased fatigue life [2], improved resistance to stress corrosion cracking [3], as well as permanent deformation of the samples [4]. LSP has been shown to be more effective than more traditional shot peening because the magnitude and depth of compression induced by LSP are greater [5]. LSP therefore has many potential applications, such as in the aerospace industry.

There are a number of LSP parameters that can be varied so as to optimize the process for different applications, e.g. power intensity, laser spot size and coverage. As part of the development of a particular laser system, it is important to understand the effect these parameters have on the induced residual stress field for various alloys and sample thicknesses. Of particular interest are the surface residual stress, the maximum compressive residual stress, as well as the depth to which the residual stresses are compressive.

In order to obtain a depth-resolved residual stress profile that can reveal this information, a number of complementary residual stress measurement techniques need to be used [6]. These include: Laboratory X-Ray Diffraction (LXRD) which can non-destructively measure near the surface (about 25 µm); Neutron Diffraction (ND) and Synchrotron XRD (SXRD) which can non-destructively measure through the entire depth of most materials (but with a larger gauge volume so the results are averaged over a greater depth); Incremental Hole Drilling (IHD) which can measure semi-destructively to an intermediate depth (1 – 2 mm), but is only accurate in samples thicker than 5.13 mm for conventional IHD strain gauge rosettes.

The objective of this study was therefore to compare the residual stress results in two LSP treated aluminium alloy 7075 (an aeronautical alloy) plates with different thicknesses investigated with a number of complementary techniques.

Methodology

Sample Preparation. Aluminium alloy 7075-T651 plate samples with dimensions 60 mm x 60 mm were prepared from a single 15 mm thick rolled plate. One side of the samples was machined to remove 1 mm followed by machining on the opposite side to respective thicknesses of 6 mm and 1.6 mm, as shown in Fig. 1a. This approach was followed so that the LSP treatment was performed on a similarly prepared surface (the one with 1 mm removed) for consistency. The bulk elastic constants were taken as E = 71.7 GPa and v = 0.33 and the corresponding diffraction elastic constants for the {311} lattice plane as: S_1 = -5.158 x 10^{-6} MPa^{-1} and $1/2S_2$ = 1.957 x 10^{-5} MPa^{-1}.

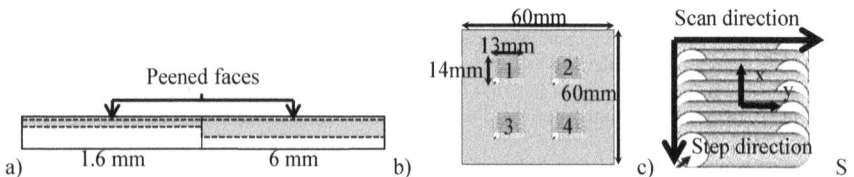

Fig. 1: a)Sample Preparation, b) Sample Geometry and LSP patches, c) raster pattern indicating the laser step and scan directions.

Laser Shock Peening. LSP was performed at the National Laser Centre of the CSIR, South Africa, using an Nd:YAG laser system with a wavelength of 1064 nm and a pulse frequency of 20 Hz. The confinement medium was flowing water and no protective coating was used. The LSP parameters were as follows: a power intensity of 3 GW/cm^2, a spot diameter of 1.5 mm and a coverage of 500 spots/cm^2. Four 13 mm x 14 mm LSP patches were applied to the sample, as shown in Fig 1b, and the LSP was performed in a raster pattern, as shown in Fig. 1c. The x- and y-directions used in subsequent sections are aligned with the step and scan directions respectively and the z-direction is normal to these, through the depth of the samples.

Laboratory X-Ray Diffraction. LXRD measurements were performed on the 6 mm thick sample (with and without LSP) using a Bruker D8 Discover at Necsa, South Africa. X-rays from

MECA SENS 2017   Materials Research Forum LLC
Materials Research Proceedings **4** (2018) 117-122   doi: http://dx.doi.org/10.21741/9781945291678-18

a Cu-Kα source with a wavelength of 1.54 Å and a beam size of 0.8 mm were used to perform the measurements in the $\sin^2\psi$ configuration. Reflections from the {311} lattice plane were measured at 16 ψ angles from -70.5° to 70.5° at a 2θ angle of approximately 78.05°. These were taken at 6 φ angles (0°, 45° and 90° and at 180° to each of these to obtain the negative ψ angles). With these measurements and the plane stress assumption, the in-plane stress tensor could be obtained. At each point, the sample was oscillated over a distance of 2 mm to improve counting statistics due to the large grain sizes as well as texture in the rolled plate samples. The measurement depth was estimated to be 26 μm using the AbsorbDX software.

Incremental Hole Drilling. IHD was performed using the SINT Restan MTS3000 Automatic Hole Drilling Machine for both sample thicknesses. Vishay type A strain gauge rosettes with a rosette diameter of 5.13 mm and nominal hole diameter of 1.8 mm were used. The holes were drilled at the centre of the LSP patches and the rosettes were aligned such that gauge 1 was aligned with the x-direction and gauge 3 with the y-direction. Each hole was drilled to a depth of 1.2 mm in steps of 20 μm. The residual stresses were calculated according to the method outlined in the ASTM E837-13 standard [7] with calibration coefficients adapted to the particular strain gauge used [8]. It should be noted that the 1.6 mm thick samples fell outside of this standard for these particular strain gauge rosettes, which were used despite this study due to availability and cost effectiveness. The residual stresses were calculated at depths from 25 – 975 μm with steps of 50 μm. Since this technique measures macro strains, the appropriate bulk elastic constants were used for the stress calculations. At least two measurements were performed for each case to ensure repeatability.

Neutron Diffraction. Monochromatic angle-dispersive ND was performed at the MPISI diffractometer at Necsa, South Africa. The neutron beam had a wavelength of 1.67 Å and both the incident and secondary slits set to 0.3 mm x 10 mm which created a matchstick shaped gauge volume. This shape was used to facilitate high depth resolution to capture the high stress gradient near the surface of the samples with as small a gauge volume as possible in the depth direction. The {311} lattice plane was measured at a diffraction angle of approximately 85.5°. Strains were measured in the x-, y- and z-directions by aligning the samples so that the direction of strain measurement bisected the incident and diffracted beams and with the long dimension parallel to the vertical surface. The gauge volume projected 0.44 mm into the depth with the sample in this orientation. Measurements were taken in the 6 mm thick sample at depths from 0.1 – 0.3 mm with steps of 0.1 mm, 0.5 – 1.5 mm with steps of 0.2 mm and from 2 – 5.5 mm with 0.5 mm steps. In the 1.6 mm sample, measurements were taken at depths of 0.1 – 0.4 mm with steps of 0.1 mm, from 0.6 – 1.2 mm with steps of 0.2 mm and from 1.3 – 1.5 mm with steps of 0.1 mm. At each depth, the samples were oscillated parallel to the surface by performing measurements at 5 locations, 0.5 mm apart. Data acquisition was done against statistical counting instead of the traditional time counting to a set maximum strain uncertainty of 50 με. Entry curves were employed to ensure coincidence between the gauge volume and the sample surface [9]. With measurements at positions close to the sample surfaces, with partially submerged gauge volumes, the 180° flip approach was performed with the two results averaged to mitigate surface aberration contributions.

The reference lattice spacing and the planar stresses were calculated at each depth according to the plane-stress assumption with the equations outlined in [10]. The elastic constants listed previously were used for the stress calculations.

Synchrotron X-Ray Diffraction. SXRD Measurements were performed at the 6-BM-A beamline at the Advanced Photon Source (APS) at the Argonne National Laboratory, USA. Energy Dispersive Diffraction (EDD) measurements were made with a hardened polychromatic beam, with a maximum energy of 285 keV, from an APS bending magnet. Two germanium detectors were used, one detector vertically offset to a diffraction angle of 5.0° and the other horizontally offset to a diffraction angle of 4.8°. This allowed simultaneous measurement of the in-plane and horizontal strain components at each depth. Due to the small diffraction angle of the detectors, the direction of strain measurement is nearly perpendicular to the primary beam. Both the beam size and the detector slits were set to 0.1 mm x 0.1 mm. With this geometry the gauge volume projects to a horizontal length of about 2.6 mm. The sample was aligned in the gauge volume such that the vertical detector measured the in-plane strain component, the horizontal measured the normal strain component and the projected length was parallel to the surface. For the 6 mm thick sample, measurements were taken at depths from 0.05 – 2.55 mm with a depth step of 0.1 mm and from 3.05 – 5.55 mm with a depth step of 0.5 mm. For the 1.6 mm thick sample, measurements were taken at depths from 0.05 – 1.55 mm with a depth step of 0.1 mm. At each depth, the samples were oscillated parallel to the surface by performing measurements at 5 locations, 0.1 mm apart. The strains were calculated by averaging the data from each location. A counting time of 40 s was used for each measurement.

To calculate strain from the EDD data, the average lattice parameter can be determined by utilising all the reflections available in the diffraction pattern. However, due to noise in the system, only the peaks of the {200} and {311} planes could be numerically fitted at every depth. Since the {200} plane is prone to inter-granular stresses, only the {311} peak positions were used to determine the lattice spacings and, hence, the strain. The reference spacing and the residual stresses at each depth were calculated in the same manner as for neutron diffraction.

Results and Discussion
Fig. 2 shows the residual stress results obtained using the various methods described for the 6mm thick sample in the x-direction, both with and without the LSP treatment. The stresses were also measured in the y-direction (omitted from Fig. 2 due to space restrictions) and the two components of stress had similar trends: a large compressive stress near the surface and tensile stress in the interior for the peened sample and a low stress- profile in the untreated sample. The magnitudes of the stresses were anisotropic being approximately 20 % more compressive in the x-direction (laser step) than in the y-direction (laser scan). This compares well to results in literature [1, 5].

The ND and Laboratory XRD investigations on the unpeened sample showed nearly zero residual stress. In the peened sample the ND and SXRD results show corresponding residual stress profiles through the entire depth of the sample. This suggests that the approach used to determine the residual stresses was sound, especially considering the different gauge volumes used.

The stresses vary rapidly in the near surface region. The LXRD (depth of 26 μm) and IHD (depth of 25 μm) results showed compressive residual stresses of approximately -200 MPa, whilst SXRD (depth of 50 μm), IHD (depth of 150 μm) and ND (depth of 160 μm) all showed stresses in excess of -350 MPa. The stress relaxation close to the surface may be attributed to the laser-material interaction at the surface causing reverse yielding in these regions.

Although IHD is typically not accurate near to surfaces, it does correspond well with the Laboratory XRD results.

These results reveal the surface residual stress, the maximum value of the compressive residual stress, as well as the depth to which the residual stresses are compressive. These could not have been attained with only one method which strongly supports the argument that a residual stress profile is best obtained using a variety of methods.

The fairly large error bars on the Laboratory XRD results are due to texture, as observed from the Debye-Scherer cones on the area detector. At certain ψ angles, the X-Ray counts were insufficient for peak fits. Further evidence of texture in the samples came from the ND and SXRD experiments.

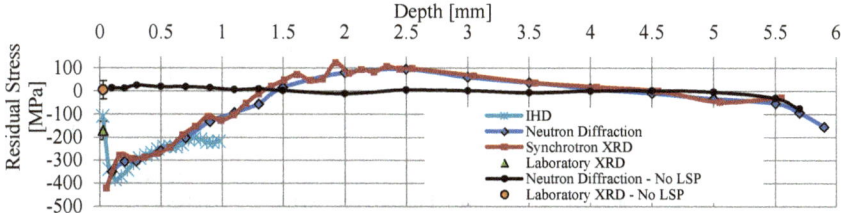

Fig. 2: Depth-resolved residuals stress results in the x-direction obtained from multiple methods on the 6 mm thick sample with and without LSP treatment.

Fig. 3 shows the residual stress results in the x-direction obtained for the 1.6mm thick sample with LSP. Again there is good correlation between the ND and SXRD results. The residual stresses are substantially less compressive than in the 6 mm thick sample. This is because there is less elastic constraint in this sample.

An offset is observed between the IHD result and the SXRD and ND below 400 μm. This can be attributed to the 1.6 mm being too thin for application of the ASTM E837 standard. It is proposed that thickness-specific calibration coefficients should be developed for samples falling outside of the standard.

Surface relaxation could not be detected with the ND and SXRD results.

It is important to note than the residual stresses are compressive through the whole depth of the sample. This means that elsewhere in the sample there will be balancing tensile stress and this needs to be considered when designing an LSP treatment – care must be taken to avoid placing a tensile residual stress in a fatigue critical location.

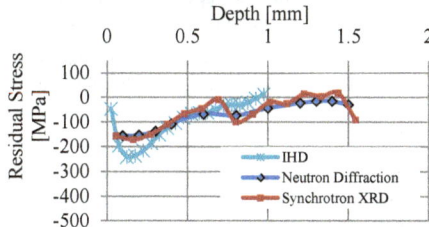

Fig. 3: Depth-resolved residuals stress results in the x-direction obtained from multiple methods on the 1.6 mm thick sample with LSP treatment.

Conclusions

Residual stresses in 6 mm and 1.6 mm thick aluminium alloy 7075 samples that had been treated with LSP have been measured using various complementary techniques. The results show high magnitude compressive residual stresses (-200 – -400 MPa) with steep gradients. Stress relaxation was shown to occur near the surface. Larger compressive residual stresses were established in the 6 mm thick sample because there was more elastic constraint provided by the underlying material volume. Additionally, the residual stresses were more compressive in the laser step direction than in the laser scan direction.

Residual stress results from the various methods employed in this study show complementary trends. Since each method has its strengths and limitations, this work showed the necessity of using multiple complementary techniques to fully describe the depth-resolved residual stress profile associated with LSP treatment.

Acknowledgements
The authors would like to acknowledge the DST-NRF Centre of Excellence in Strong Materials (CoE-SM) for their generous financial support. They would also like to acknowledge the vital support of the South African Centre for Scientific and Industrial Research National Laser Centre's (CSIR NLC) Rental Pool Program (RPP), funded by the Department of Science and Technology (DST).

In addition, this work is based on the research supported in part by the National Research Foundation (NRF) of South Africa Incentive Funding for Rated Researchers (IFRR) for Prof. Claudia Polese and Equipment-Related Travel and Training Grant (ERTTG) (Grant Numbers: 109200 and 115195).

Opinions, findings and conclusions or recommendations expressed in this work are those of the authors and are not necessarily to be attributed to the CoE-SM or to the NRF.

References

[1] K. Ding and L. Ye, Laser Shock Peening Performance and Process Simulation, Woodhead Publishing Limited, Cornwall, 2006. https://doi.org/10.1201/9781439823620

[2] T. Adachi et al., Effect of laser peening on fatigue properties for aircraft structure parts, in: International Conference on Shot Peening, 2008.

[3] Y. Sano, M. Obata, T. Kubo, N. Mukai, M. Yoda, K. Masaki and Y. Ochi, Retardation of crack initiation and growth in austenitic stainless steels, Mater. Sci. Eng. A, 417 (2006) 334-340. https://doi.org/10.1016/j.msea.2005.11.017

[4] Q. Yu, Z. Dong, R. Miao, X. Deng and L. Chen, Bending deformation of laser peened aluminium alloy with uniform rectangular spots, Mater. Sci. Technol. 32 (2016) 9-14. https://doi.org/10.1179/1743284715Y.0000000057

[5] C.A. Rodopoulos, J.S. Romere, S.A. Curtis, E.R. de los Rios and P. Peyre, Effect of controlled shot peening and laser shock peening on the fatigue performance of 2024-T351 aluminum alloy, J. Mater. Eng. Perform. 12(4) (2003) 414-419. https://doi.org/10.1361/105994903770342944

[6] N.S. Rossini, M. Dassisti, K.Y. Benyounis and A.G. Olabi, Methods of measuring residual stresses in components, Mater. Des., 32 (2012) 572-588. https://doi.org/10.1016/j.matdes.2011.08.022

[7] ASTM, ASTM E837 – 13a Standard test method for determining residual stresses by the hole-drilling strain-gage method, ASTM International, West Conshohocken, 2013.

[8] E. Valentini, A. Benincasa and L. Bertelli, Residual stress calculation using the incremental hole-drilling method with eccentric holes: the eval software, in: Associazione Italiana Per L'analisi Delle Sollecitazioni 43° Convegno Nazionale, Rimini, 2014.

[9] P.C. Brand and H.J. Prask, New methods for the alignment of instrumentation for residual-stress measurements by means of neutron diffraction, J. Appl. Crystallogr. 27 (1994) 164-176. https://doi.org/10.1107/S0021889893007605

[10] A.M. Venter, D. Marais and V. Luzin, Benchmarking studies of the MPISI Material Science Diffractometer at SAFARI-1, Mat. Res. Proc. 2 (2016) 413-418.

MECA SENS 2017
Materials Research Proceedings 4 (2018) 123-128

Materials Research Forum LLC

doi: http://dx.doi.org/10.21741/9781945291678-19

Depth-Resolved Strain Investigation of Plasma Sprayed Hydroxyapatite Coatings Exposed to Simulated Body Fluid

T. Ntsoane[1,2,a*], C. Theron[2,b], M. Topic[3,c], M. Härting[4,d] and R. Heimann[5,e]

[1]Research and Development Division, Necsa SOC Limited, Pretoria, South Africa

[2]Department of Physics, University of Pretoria, Pretoria, South Africa

[3]Materials Research Department, iThemba LABS/National Research Foundation, Cape Town, South Africa

[4]Department of Physics, University of Cape Town, Cape Town, South Africa

[5]Am Stadtpark 2A, D-02826 Görlitz, Germany

[a]tshepo.ntsoane@necsa.co.za, [b]chris.theron@up.ac.za, [c]mtopic@tlabs.ac.za, [d]margit.harting@uct.ac.za, [e]robert.heimann@ocean-gate.de

Keywords: Plasma-Sprayed Hydroxyapatite Coatings, In-Vitro Investigation, High-Energy Diffraction, Residual Stress

Abstract. The influence of exposure to simulated body fluid (SBF) on plasma sprayed hydroxyapatite (HAp) coatings on medical grade Ti6Al4V samples has been investigated. Through-thickness residual strain investigations of HAp coatings deposited on flat substrate surfaces incubated for 7, 28 and 56 days were performed using high-energy synchrotron diffraction techniques. In the as-sprayed condition, the results show the top half of the HAp coating to be under compression with the maximum around the near-surface region, relaxing with depth below the surface reaching a strain-free point around the coating thickness midpoint. On the contrary, the remainder of the coating is under tension increasing with further depth; the maximum tension is observed near the coating-substrate interface region. Upon immersion in SBF, both the slope of the normal strain components ε_{11} and ε_{33} relax, with the former experiencing a change in slope before saturating after 7 days; the highest change was observed within the first week of incubation.

Introduction

The second-generation biomaterial hydroxyapatite (HAp) has been extensively studied as a candidate material in biomedical applications due to its similarity to the mineral component of bone. These include filling of bone cavities [1] and medical implant coatings for improved biological fixation [2] amongst others. The poor mechanical properties of the material, however, limit its bulk utilisation in load-bearing applications. To overcome this limitation, the material is applied as a coating on metallic substrates such as Ti, Ti alloys and CoCrMo, combining the excellent mechanical properties of the metal with the osseoconductive ability of the coating [3]. With the plethora of coating techniques available for deposition [4], thermal spraying is still the method of choice. Although successfully utilised at an industrial scale (see, for example [5]), the high plasma temperature together with the cold substrate surface that the droplet impinges on, generally results in thermal decomposition of the HAp powder and rapid cooling on the substrate respectively. This leads to the introduction of undesirable thermal decomposition products [6], such as tricalcium phosphate (TCP), tetracalcium phosphate (TTCP), and sometimes calcium oxide as well as a reduced crystallinity [7]. These products are known to be susceptible to dissolution in simulated body fluids [8] and thus together with the strains and stresses generated

MECA SENS 2017
Materials Research Forum LLC
Materials Research Proceedings 4 (2018) 123-128
doi: http://dx.doi.org/10.21741/9781945291678-19

as a result of differential thermal mismatch (CTE) and quenching of the droplet, may compromise the mechanical stability and integrity of the coating. Although extensive investigation of the effect of incubation of HAp coatings in simulated body fluid have been carried out by many research groups around the world [6], the bulk of their work focused on the near-surface region coating, i.e. the region in immediate contact with living tissue. The present study is an extension of the author's previous work [9] on through-thickness investigation of HAp coating in the as-sprayed condition.

Materials and methods
Sample preparation: Hydroxyapatite powder (CAPTAL 90, batch P215, Plasma Biotal Limited, Tideswell, Derbyshire, UK) with size distribution of 120 ±20 μm was plasma-sprayed onto flat discs of 20 mm diameter medical grade Ti6Al4V alloy substrate supplied by Biomaterials Limited, North Yorkshire, UK. Details of deposition and spray parameters have been reported elsewhere [9]. Subsequent to spraying, the samples were incubated in simulated body fluid for 7, 28 and 56 days to mimic the physiological environment. Sample incubation was carried out in a revised simulated body fluid (rSBF) based on Kokubo's formulation [10]. The solution had an ionic concentration similar to the human blood plasma but without proteins and enzymes. The temperature and *pH* of the solution during incubation experiment were kept at 36°C and *7.4*, respectively. Subsequent to immersion, slices of approximately 5 mm thick were cut for investigation. Factors considered in determining the optimum slice length are reported elsewhere [9].

Through-thickness characterisation of HAp coating: Angular dispersive diffraction measurements utilising the high–energy synchrotron radiation, 100 keV (wavelength λ = 0.12331 Å), at the Advanced Photon Source's X-ray Operation and Research 6-ID-D beamline at Argonne National Laboratory , USA was used for the experiments. Experimental details and measurement procedure have been reported elsewhere [9]. Measurements were done in transmission geometry using a 35(V) x 400 (H) μm^2 beam and for one azimuth orientation hence the full strain tensor was not measured. The analysis of the data for phase composition and strain was done using TOPAS [11] and the traditional one-dimensional method [12] respectively. The error calculation in the latter was based on the standard deviation of the fit assuming a Gaussian distribution.

Cross-section microstructure examination of the as-sprayed and sample subjected to immersion in simulated body fluid was done using scanning electron microscope. For better quality micrographs, the samples were metallographically prepared and images obtained in secondary electron (SE) mode.

Results and discussion
Phase analysis: Fig.1 show the superposed diffraction patterns of the as-sprayed and the sample immersed the longest (56 days) as well as the corresponding volume fractions of the starting HAp phase and main thermal product (TTCP) collected at different depths below the coating surface; the bottom patterns in the former represent the shallow depths probed in this geometry. The last top diffraction pattern(s) in the figures corresponds to the Ti alloy indicating that probing extended beyond the coating-substrate interface. The high temperature induced thermal products TTCP, TCP, and CaO can be seen through-out the as-sprayed coating, see Fig. 1a. Upon immersion these phases dissolve, with the latter being first to disappear. After 56 days of immersion, CaO has almost completely disappeared while TTCP and TCP only start appearing deeper in the coating see Fig. 1b.

MECA SENS 2017   Materials Research Forum LLC
Materials Research Proceedings **4** (2018) 123-128   doi: http://dx.doi.org/10.21741/9781945291678-19

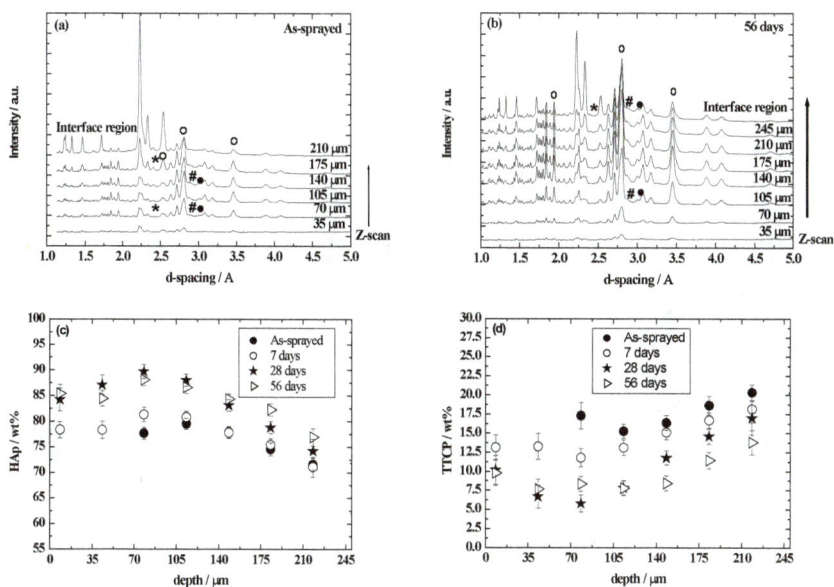

Fig. 1: Through-thickness diffraction patterns of HAp coatings: (a) as-sprayed, (b) 56 days immersed samples showing the presence of HAp (o), TTCP (#), TCP (•) and CaO (*) and the corresponding volume fractions: (c) HAp and (d) TTCP.

The corresponding volume fraction for the HAp phase as well as the main thermal product, TTCP for all immersion periods are shown in Figs. 1c and d. For the as-sprayed condition, HAp increases linearly with depth from ~75 wt.% at the near-interface region to a maximum of 80 wt.% at around 105 μm before decreasing to ~77 wt.% near the coating surface. The observed reduction in HAp closer to the coating surface might be attributed to slower cooling rates towards the outside of the layer due to the already elevated temperature and the lower thermal conductivity of HAp. Upon immersion the amount of HAp increases gradually reaching a maximum around 28 days of immersion. The observed relative increase is attributed to the dissolution of thermal products from the coating into the solution, leaving behind a more stable HAp phase. An opposite trend is observed for the thermal product TTCP in the as-sprayed condition, decreasing from ~22.75 wt.% at the interface region reaching a minimum of ~15 wt.% before increasing near the coating surface. It decreases further upon immersion.

Microstructure: Fig. 2 shows the SEM micrographs of the as-sprayed and SBF immersed coatings. In general the micrographs show a laminar-based structure indicative of the splat-based nature of the coating. The interface region of the as-sprayed coating shows a slightly higher number of partially and/ or unmolten particles, Fig. 2a. The fine-grained equiaxial microstructure revealed by microscopy is an indication of low heat removal from the deposited splats. This is to be expected given the already elevated substrate temperature due to heat flux from the plasma jet and the already deposited HAp splats. Additionally, the presence of high concentration of splat boundaries, fine boundaries and gaps/pores resulting from poor adhesion between the splats can be seen, as well as cracks resulting from in-plane residual stresses. Upon immersion the laminar

MECA SENS 2017   Materials Research Forum LLC
Materials Research Proceedings **4** (2018) 123-128   doi: http://dx.doi.org/10.21741/9781945291678-19

structure is replaced by very fine grains, as well as nano-sized pores resulting in an increased coating porosity. The latter connects with one another resulting in a 3D network of dissolution channels which facilitate dissolution deeper in the coating. With further immersion , an apatite-like precipitate layer (of thickness ~20 μm) is formed on the coating surface, Fig. 2b. Closer examination of the micrograph reveals that the precipitate is not only limited to the coating surface but extends into the bulk, thereby filling some pores.

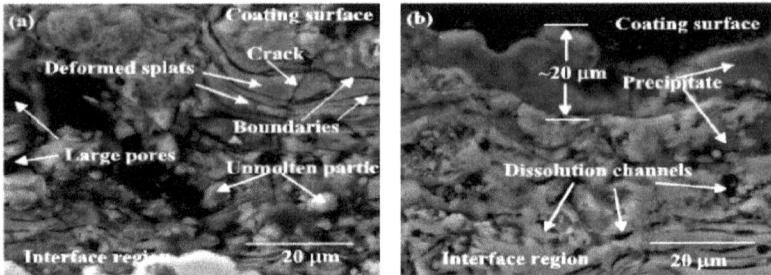

Fig. 2: Cross-section morphology of HAp coating for different immersion times: a) as-sprayed and b) 56 days. The latter indicates presence of a layer of apatite-like precipitate.

Residual strain: Figs. 3a and 3b show the through-thickness normal strain components ε_{11} and ε_{33} of the coatings for the different immersion periods respectively. The values are calculated from the 213 reflections of the main phase HAp with the stress-free reference d-spacing d_0 determined from the powder prepared from sprayed coating flakes. This was done to ensure that the observed changes were due to residual stress and not chemistry changes induced by spraying. The observed plots show the strain in the as-sprayed condition to be generally small with the normal component ε_{11} under tension in the first half of the coating. It relaxes linearly with depth below the coating surface reaching a strain-free point around the coating midpoint from which a change in strain state is observed. An opposite trend is observed for the normal component ε_{33}. Dissolution behaviour corresponding to the slope of the fits to the data points, $\Delta\varepsilon_{ij}$ as a function of immersion time are summarised in Figs. 3c and 3d. From the figures the slope of the strain component ε_{11}, $\Delta\varepsilon_{11}$ relaxes to zero before turning positive after 7 days, remaining constant with further immersion. On the other hand, the slope for the strain component ε_{33}, $\Delta\varepsilon_{33}$ relaxes linearly with depth to zero after 56 days of immersion in SBF. The observed $\Delta\varepsilon_{11}$ behaviour (in this study) is similar to the one for the coating deposited on a cylindrical rod geometry, however this is not the case for $\Delta\varepsilon_{33}$. The latter showed significant relaxation from positive to zero in the first 7 days before stabilizing with further immersion.The observed relaxation and/or change in strain state can be attributed to dissolution of the thermal products and the amorphous content in the coating leaving behind a more stable crystalline HAp, as well as the formation of apatite-like precipitate.

The observed as-sprayed distribution trend is consistent with the findings of Cofino *et al.* [13]. A combination of the high quenching effect upon droplet impact on cooler substrate and slightly higher thermal expansion coefficient of the coating as compared to the substrate are attributed to the higher coating tension at the interface region. The observed strain relaxation and change in strain state i.e. from compression to tension, upon immersion is attributed to the dissolution of thermal products and precipitate layer formed. The observed change in strain/stress state upon SBF immersion is consistent with Nimkerdphol *et al.* [14].

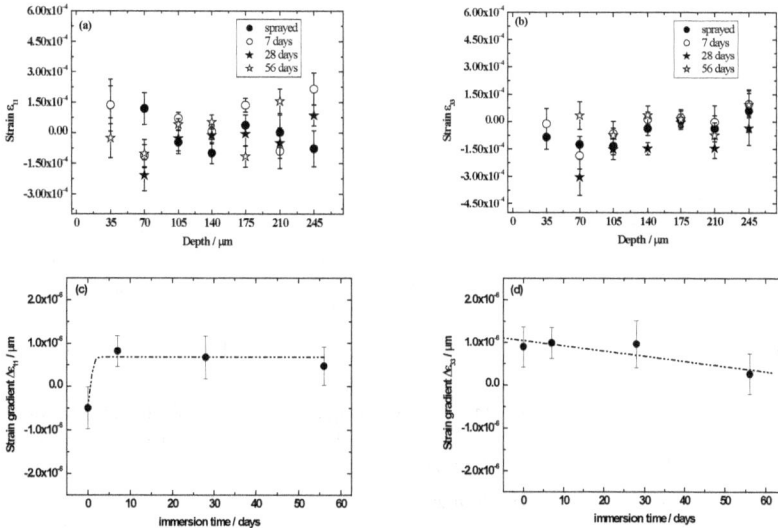

Fig. 3: Variation of strain ε_{ij} and strain gradient $\Delta\varepsilon_{ij}$ with depth and immersion time: a) ε_{11}, b) ε_{33}, c) $\Delta\varepsilon_{11}$, and d) $\Delta\varepsilon_{33}$. (The dotted lines are guides to the eye)

Summary

The effects of simulated body fluids on air-plasma sprayed HAp coatings deposited on flat geometry substrate were investigated employing synchrotron radiation and the following were revealed:

a) Phase identification results showed thermal products to be present throughout the coating of the as-sprayed sample with CaO amongst the first to dissolve upon immersion in SBF. Quantitative analysis showed HAp to be increasing with immersion while TTCP show an oppotite trend.

b) Both the two strain component slopes $\Delta\varepsilon_{11}$ and $\Delta\varepsilon_{33}$ relax upon immersion in SBF with the former showing significant changes within the first 7 days of immersion and stabilising with further immersion time. The latter relaxes linearly with immersion reaching strain free after 56 days.

Acknowledgements

Use of the Advanced Photon Source (APS) at Argonne National Laboratory (ANL) was supported by the U.S. DOE under Contract No. DE-AC02-06CH11357. The authors are indebted to Mrs Margitta Hengst, Department of Mineralogy, Technische Universität Bergakademie Freiberg, Germany for sample preparation, Drs Douglas Robinson and Jonathan Almer, ANL/APS for assisting with the experimental set-up and data acquisition and analysis, respectively as well as Mr. Ryno van der Merwe, for microscopy analysis. The research work was sponsored by the German Federal Ministry of Education and Research (BMBF) and the National Research Foundation of the Republic of South Africa within the research project "Characterisation and determination of residual stress in bioactive coatings and layered structure" (Project code 39.6.G0B.6.A).

References

[1] T. Yamamoto, T. Onga, T. Marui and K. Mizuno, Use of hydroxyapatite to fill cavities after excision of benign bone tumours, J. Bone Joint Surg. (Br.) 82B (2000) 1117.

[2] R.G.T. Geesink, K. de Groot and C.P.A.T. Klein, Chemical implant fixation using hydroxyl-apatite coatings The development of a human total hip prosthesis for chemical fixation to bone using hydroxyl-apatite coatings on titanium substrates, Clin. Orthop. Relat. Res. 225 (1987) 147.

[3] K. de Groot, R. Geesink, C.P.A.T. Klein and P. Serekian, Plasma sprayed coatings of hydroxylapatite, J. Biomed. Mater. Res. 21 (1987) 1375. https://doi.org/10.1002/jbm.820211203

[4] R.B. Heimann and H.D. Lehmann, Bioceramic Coatings for Medical Implants Bioceramic Coatings for Medical Implants. Trends and Techniques. Wiley-VCH, Weinheim, Germany. (2015) 467 pp. 113. https://doi.org/10.1002/9783527682294

[5] R.B. Heimann, Thermal spraying of biomaterials, Surf. Coat. Technol., 201 (2006) 2012. https://doi.org/10.1016/j.surfcoat.2006.04.052

[6] M. Topić, T. Ntsoane, T. Hüttel and R.B. Heimann, Microstructural characterisation and stress determination in as-plasma sprayed and incubated bioconductive hydroxyapatite coatings, Surf. Coat. Technol. 201(6) (2006) 3633. https://doi.org/10.1016/j.surfcoat.2006.08.139

[7] K.A. Gross, C.C. Berndt and H. Herman, Amorphous phase formation in plasma-sprayed hydroxyapatite coatings, J. Biomed. Mater. Res. 39 (1998) 407. https://doi.org/10.1002/(SICI)1097-4636(19980305)39:3%3C407::AID-JBM9%3E3.0.CO;2-N

[8] P. Ducheyne, S. Radin and L. King, The effect of calcium phosphate ceramic composition and structure on in vitro behavior. I. Dissolution, J. Biomed. Mater. Res. 27 (1993) 25. https://doi.org/10.1002/jbm.820270105

[9] TP Ntsoane, M. Topic, M. Härting, R.B. Heimann and C. Theron, Spatial and depth-resolved studies of air plasma-sprayed hydroxyapatite coatings by means of diffraction techniques: Part I, Surf. Coat. Technol. 294 (2016) 153-183. https://doi.org/10.1016/j.surfcoat.2016.03.045

[10] T. Kokubo, H. Kushitani, S. Sakka, T. Kitsugi and T. Yamamuro, Solutions able to reproduce in vivo surface-structure changes in bioactive glass-ceramic A-W, J. Biomed. Mater. Res. 24 (1990) 721-734. https://doi.org/10.1002/jbm.820240607

[11] TOPAS 4.2 2009 Bruker AXS.

[12] J. Almer and U. Lienert, Unpublished document for the Neutron/Synchrotron Summer School, Argonne National Laboratory, 2001.

[13] B. Cofino, P. Fogarassy, P. Millet and A. Lodino, Thermal residual stresses near the interface between plasma-sprayed hydroxyapatite coating and titanium substrate: Finite element analysis and synchrotron radiation measurements, J. Biomed. Mater. Res. (2004) 1-70. https://doi.org/10.1002/jbm.a.30044

[14] A.R. Nimkerdphol, Y. Otsuka and Y. Mutoh, Effect of dissolution/precipitation on the residual stress redistribution of plasma-sprayed hydroxyapatite coating on titanium substrate in simulated body fluid (SBF), J. Mech. Behav. Biomed. Mater. 36 (2014) 98-109. https://doi.org/10.1016/j.jmbbm.2014.04.007

MECA SENS 2017   Materials Research Forum LLC
Materials Research Proceedings **4** (2018) 129-134   doi: http://dx.doi.org/10.21741/9781945291678-20

Stress Profiling in Cold-Spray Coatings by Different Experimental Techniques: Neutron Diffraction, X-Ray Diffraction and Slitting Method

V. Luzin[1,a*], K. Spencer[2,b], M.R. Hill[3,c], T. Wei[1,d], M. Law[1,e]
and T. Gnäupel-Herold[4,f]

[1]Australian Nuclear Science & Technology Organisation, Lucas Heights, NSW 2234, Australia

[2]School of Mechanical and Mining Engineering, The University of Queensland, QLD 4072, Australia

[3]Department of Mechanical and Aerospace Engineering, University of California, Davis, CA 95616, USA

[4]NIST Center for Neutron Research, National Institute of Standards and Technology, Gaithersburg, MD 20899, USA

[a]vladimir.luzin@ansto.gov.au, [b]krspencer@gmail.com, [c]mrhill@ucdavis.edu, [d]tao.wei@ansto.gov.au, [e]michael.law@ ansto.gov.au, [f]thomas.gnaeupel-herold@nist.gov

Keywords: Residual Stress, Coatings, Cold Spray, Neutron Diffraction

Abstract. The residual stress profiles in Cu and Al coatings sprayed using kinetic metallization to thickness of ~2 mm have been studied. Due to specific parameters of the cold-spray process and particular combination of materials, coatings and substrates, the residual stresses are low with magnitudes of the order of a few tens of MPa. This poses challenges on accuracy and resolution when measuring through-thickness stress distributions. Three experimental techniques - neutron diffraction, X-ray diffraction and a slitting method - were used to measure through-thickness stress distributions in the substrate-coating systems. All three techniques demonstrated acceptable accuracy and resolutions suitable for analyzing stress profiles. Advantages and disadvantages of each technique are discussed.

Introduction

Coatings of many different materials and thicknesses find their use in various surface enhancement applications such as wear resistance, corrosion protection, insulation, etc. They can be deposited on surfaces of engineering components by a number of techniques including thermal and cold spray. These techniques are energetic processes that generally induce residual stresses either through thermal effects or kinetic impact. The residual stresses can be detrimental for the coating's mechanical integrity or functional performance, therefore stress control or mitigation is usually required.

For coatings with thickness of a few millimeters, several stress measurement techniques can be used. Neutron diffraction stress measurement has proven to be a useful method for thick [1] and thin coatings [2] owing to some advantages: It is non-destructive and it can provide the required high resolution (down to 0.2 mm); No special sample preparation (e.g. cutting and polishing, as for X-rays) is required: Measurements can be done in a reasonable time (minutes per datum) and with high accuracy (uncertainty can be better than 5 MPa). The slitting method, despite it being a destructive technique, has also been proven to be an efficient method for stress measurements due to the high spatial resolution and accuracy it provides [3]. The laboratory X-ray diffraction technique is not commonly used for through-thickness stress measurement in coatings, nor is it associated with high spatial resolution as required for the investigation of

MECA SENS 2017   Materials Research Forum LLC
Materials Research Proceedings 4 (2018) 129-134   doi: http://dx.doi.org/10.21741/9781945291678-20

coatings. Some advances in the technique [4] and demonstration of its ability in application to thin (~1 mm) metal sheets [5] open opportunities for a new application of this technique. Since the laboratory X-ray technique is most accessible, it gives certain advantages to this technique.

In this work we report on an experimental study of the residual stress analysis in cold-spray coatings made with the three techniques mentioned.

Sample production
Two coatings were sprayed using the kinetic metallization (KM) cold-spray technique with a simple convergent barrel nozzle at a nozzle standoff distance of 12 mm and a nozzle traverse speed of 50 mm/s. The gas temperature-pressure conditions, optimized for deposition efficiency and reported in Table 1 were also used to estimate the exit velocity of the particles.

The deposition of powder on flat copper coupons (30×30 mm^2, 3 mm thickness) resulted in coatings with a thickness of ≈ 2.1 mm for both materials.

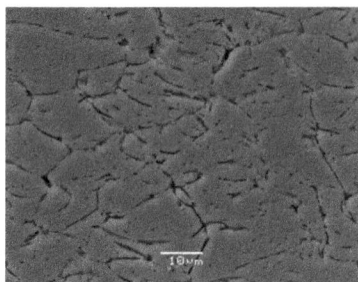

Fig. 1: Microstructure of the Al coating.

Table 1. Sample spraying conditions and parameters.

Powder Material	Average Particle Size [μm]	Driving Pressure [kPa]	Nozzle Temperature [°C]	Powder Feed Rate [g/min]	Estimated Particle Exit Velocity [m/s]
Al	15	620	140	15	585
Cu	6	620	200	15	645

Sample characterization
(i) For measurements of the Young's modulus, rectangular specimens were extracted from the bulk of the coating with approximate dimensions $30 \times 5 \times 2$ mm^3. The Young's modulus E was determined using the impulse excitation technique (ASTM standard E1876) through acoustic measurements of the normal frequency and the sample dimensions. The accuracy of this method was better than 1 %. The same samples were used for density ρ measurements (Table 2).

(ii) In order to determine the phase composition, the full neutron diffraction patterns were measured (using the very same bar samples used for the Young's modulus measurement) in the 2θ range of $10° - 160°$ at a wavelength of 1.62 Å using high-resolution powder diffractometer ECHIDNA at the ANSTO OPAL research reactor [6]. The volume fractions of the phases were extracted through a refinement procedure (Table 2).

Table 2. Coating's material characterization: Young's modulus, density and phase composition.

ID	E [GPa]	% of bulk	ρ [g/mm^3]	% of bulk	Phase composition, vol %
Al	49.4 ± 0.2	69.6 ± 0.3	2.528 ± 0.016	93.6 ± 0.6	Al 100 %
Cu	104.0 ± 0.5	83.9 ± 0.6	8.670 ± 0.044	98.5 ± 0.5	Cu95 %+Cu$_2$O 5 %

Neutron residual stress measurements
The neutron diffraction residual stresses measurements were carried out at the NIST Center for Neutron Research using the BT8 residual stress diffractometer [7]. In correlation to the sample thicknesses a $0.5 \times 0.5 \times 18$ mm^3 sized gauge volume was used. For the most optimal definition of the gauge volume a 90˚ ($2\theta_B \sim 90°$) diffraction geometry was employed by setting the take-off angle $2\theta_M$ of the Si(311) monochromator and the wavelength λ accordingly (Table 3). Through-

MECA SENS 2017   Materials Research Forum LLC
Materials Research Proceedings 4 (2018) 129-134   doi: http://dx.doi.org/10.21741/9781945291678-20

thickness measurements were done in locations within the substrate and coating with 0.3 mm spacing between points. For each measurement point d-spacings were measured in the two principal directions, normal to the surface and in-plane. The d-spacings of the (311) reflections were measured with sufficient statistics to provide at least 5×10^{-5} uncertainty in the strain. Since Cu is a stronger neutron scatterer, 10 minutes per measurement point (t) were adequate for Cu, while 90 minutes were required for Al. Assuming a balanced biaxial plane-stress state, the stress values were calculated according to the procedure [1] from the measured d-spacings and the diffraction elastic constants estimated by the self-consistent Kröner method [9] (Table 3).

Table 3. Neutron instrument setting and material constants for the measured reflections.

	d [Å]	$2\theta_M$	λ [Å]	$2\theta_B$	S_1 [TPa^{-1}]	$\frac{1}{2}S_2$ [TPa^{-1}]	t [min]
Cu(311)	1.10	70.0°	1.55	89.7°	-3.38	12.58	10
Al(311)	1.22	79.8°	1.72	89.5°	-5.16	19.57	90

X-ray diffraction measurements
On completion of the neutron diffraction measurements, the samples were cut in halves and the surface of the cross-sectional cut investigated with X-ray measurements (after appropriate surface preparation). The measurement technique used in the high spatial resolution measurements was described previously [4]. It utilizes a very narrow vertical beam, 0.05 mm in our case, a combination of Ω- and Ψ-angles, as well as sample rotation applied in such a way that ensures that the X-ray beam projection remains parallel to the surface/interface, therefore representing the same through-thickness depth, and does not exceed the desired spatial resolution, 0.1 mm in our case.

The Cu-tube K-α radiation was used to measure the Cu(420) and Al(511/333) reflections. Although measurement time with such a small beam is an order of magnitude longer (t = 120 seconds for each orientation), many through-thickness locations could be measured in the available beamtime. The experimental details of the X-ray measurements are summarised in Table 4.

The primary interest was on locations immediately adjacent to the interface, thus posing a challenge for the neutron technique. Thus, 6 points in each coating were measured in steps of 0.1 mm with the first point located 0.1 mm away from the interface. The average uncertainty of X-ray stress values was better than 5 MPa, a prerequisite to resolve subtle stress variations.

An attempt was also made to measure stresses in the Cu substrate, but due to the large grains, statistical variations were too great to provide a reliable result notwithstanding measurements taken for two Cu reflections, Cu(420) and Cu(331) and used in combination. Therefore, this data was omitted from the publication.

Table 4. X-ray instrument setting and material constants for the measured reflections.

	d [Å]	Tube	λ [Å]	$2\theta_B$	S_1 [TPa^{-1}]	$\frac{1}{2}S_2$ [TPa^{-1}]	t [sec]
Cu (420)	0.8061	Cu	1.54433	145°	-2.87	11.01	120
Al (511/333)	0.7793	Cu	1.54433	163°	-5.15	19.69	120

Slitting measurements
For the slitting method stress measurements [3], 5 mm thick slices were cut from each of the X-ray samples to have $5 \times 5 \times 30$ mm^3 bars available for these tests. (Through-thickness dimension size of 5.3 mm is composed of the 3.2 mm thick Cu substrate and a 2.1 mm thick coating). A strain gauge attached to the surface opposite to the coated surface was used to measure strain changes when a slit was cut through the sample thickness by wire EDM. Cut depth increments

MECA SENS 2017   Materials Research Forum LLC
Materials Research Proceedings **4** (2018) 129-134   doi: http://dx.doi.org/10.21741/9781945291678-20

were 0.1 - 0.2 mm, with a final cut depth of about 4.5 mm from the top surface. This allowed for reconstruction of the stresses in the coatings and most of the substrate thicknesses.

The reconstruction of the stress profiles from the strain gauge data was done in two different ways: The unit-pulse analysis approach using a Tikhonov regularisation to reconstruct stress profiles of different degrees of smoothness and statistical robustness; Eigenstrain analysis approach where the stress profile solution is sought within *a priori* determined class of functions (e.g. polynomial of nth order) as a best fit to the experimental data. A comparison of the two approaches is given if Fig. 2 for the Cu/Cu system. The second approach (eigenstrain, order = 1) in the case of sprayed coating has certain advantages since it is known that: (i) the stress profile in the substrate is a linear function (elastic bending response to bring about force and bending moment balances); (ii) the stress profile in the coating typically is a very smooth function, very close to a linear function (as demonstrated below by modelling results). In fact, this type of solution is the only one that can provide a reasonable result in the most difficult case, i.e. the Al/Cu system.

FEM modelling

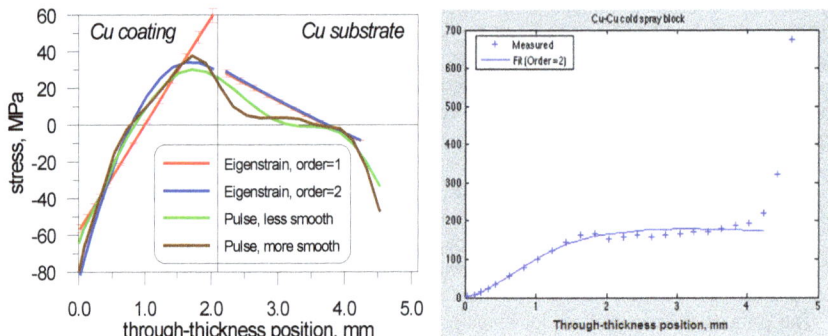

Fig. 2: Slitting method analysis of the Cu/Cu system: (left) several solutions of the reconstructed stress and (right) a corresponding strain function (μstrain vertical axis) fitting the strain gauge experimental data for one of the solutions. The results are provided in the coordinate system of the EDM cut: zero point corresponds to the top coating surface.

The most critical part of the stress analysis in coatings is the geometrical conditions of the investigated samples and the relationship between measured stresses. In reality, the measured strains/stresses are not exactly the same (Fig. 3): (i) The neutron measurements were done in the central part of the sample, away from the edges, ensuring that the original equi-biaxial (typical for most coatings) stress state is measured; (ii) For the X-ray technique measurements are made on a cut, so the stress state is not equi-biaxial anymore and the normal stress component is eliminated by the presence of a free surface. One in-plane component remains, but may be altered by the cut; (iii) For convenience of the slitting method, the original sample was altered even further. The second cut to produce a 5 mm bar further reduces constraints from the other parts of the half-sample, thus changing the stress state again.

To address the issue of the different sample geometries and relationships between stress states the Chill modelling approach [9] was applied on a generalized bi-metal two-layer plate sample to emulate the coating/substrate system. The system was 5 mm thick Al (E = 70 GPa) and 3 mm thick steel (E = 210 GPa) with plate dimensions 100×100 mm^2. Instead of a stress profile

MECA SENS 2017   Materials Research Forum LLC
Materials Research Proceedings 4 (2018) 129-134   doi: http://dx.doi.org/10.21741/9781945291678-20

generated by a deposition process, a stress profile was induced here thermally through a combination of different coefficients of thermal expansion and temperature increments.

Using ABAQUS, the two consecutive cuts were simulated by tracing the stress relaxations and stress profiles at each step. The results of the simulation are given in Fig. 4. Although some geometry effects are visible in the simulation as distortions of the stress profile from the ideal (theoretically linear), the general trend of the stress distribution demonstrates that a stress relaxation effect is observable, but not at a significant level. Some 80 – 90 % of the original stress magnitude is preserved in the bar or edge sample. Although, some corrections (a factor) can be applied to account for the partial stress relaxation of the remaining in-plane component, but in our case this effect was considered negligible.

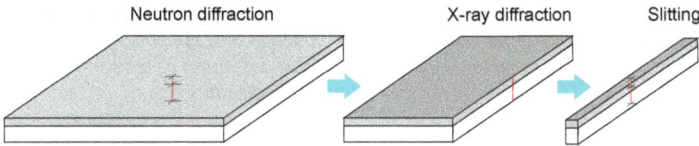

Fig. 3: Three different sample geometries used for the investigation with the three techniques.

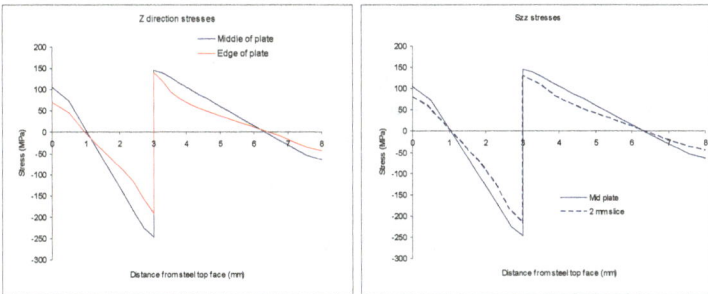

Fig. 4: Relaxation of the stress on the sample edge (left) and in the thin bar (right) in comparison with the stress in the middle of the uncut sample.

Results and discussion

The results of the three experimental methods are combined in Fig. 5. Overall, all three methods demonstrated an ability to resolve the stresses with good accuracy, < 5 – 10 MPa, even in the case of very subtle stress distribution in the Al/Cu system. It is therefore confirmed that any of the techniques is suitable for stress measurements in coatings.

While neutrons and X-rays are in very good agreement, the slitting methods results have some visible deviation from the diffraction results, as is seen most vividly in the Cu/Cu system. Upon reviewing, there are at least two objective reasons for results of the slitting method to differ. Both of them are linked to the fact that the top surface is not smooth, but exhibits a significant variation of ±0.4 mm from the average, as measured from minimum and maximum values (Fig. 5). Due to this, depending on the exact location of the EDM cut, the effective coating thickness can vary within a range, and strain relaxation readings will be sensitive to this effect. Secondly, for the stress reconstruction procedure, the coating and substrate thicknesses are parameters of the elastic model and therefore, if average (vs local) thickness is used, this will also impact the final stress values.

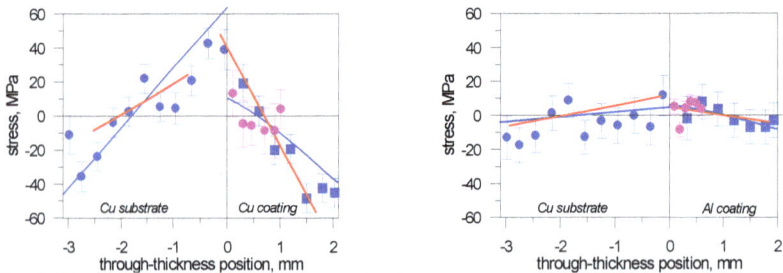

*Fig. 5: The results of stress determination by the three methods for the two coating systems,
Cu/Cu (left) and Al/Cu (right). Neutron data are shown in blue symbols with Tsui & Clyne
model [10] (blue line) fitting to the datasets. X-ray data are shown as magenta symbols. The
red line is for the slitting method results.*

In the cases of neutrons and X-rays, the measurement and stress calculation procedures rely on average parameters leading to more robust results with better agreement.

*Fig. 6: Variation of the Al coating
thickness profile: $t_{min} = 1.6$ mm,
$t_{max} = 2.8$ mm, $t_{ave} = 2.0$ mm.*

References

[1] V. Luzin, A. Valarezo and S. Sampath, Through-thickness Residual Stress Measurement in Metal and Ceramic Spray Coatings by Neutron Diffraction, Mater. Sci. Forum 571-572 (2008) 315-320. https://doi.org/10.4028/www.scientific.net/MSF.571-572.315

[2] V. Luzin, A. Vackel, A. Valarezo and S. Sampath, Neutron Through-Thickness Stress Measurements in Coatings with High Spatial Resolution, Mater. Sci. Forum 905 (2017) 165-173. https://doi.org/10.4028/www.scientific.net/MSF.905.165

[3] M.R. Hill, The Slitting Method, in: Practical Residual Stress Measurement Methods, John Wiley & Sons, Ltd, 2013, pp. 89-108. https://doi.org/10.1002/9781118402832.ch4

[4] T. Gnaupel-Herold, Formalism for the determination of intermediate stress gradients using X-ray diffraction, J. Appl. Crystallogr. 42 (2009) 192-197. https://doi.org/10.1107/S0021889809004300

[5] T. Gnaeupel-Herold, T. Foecke, M. Iadicola and S. Banovic, in: SAE International, 2005.

[6] K.-D. Liss, B. Hunter, M. Hagen, T. Noakes and S. Kennedy, Echidna – the new high-resolution powder diffractometer being built at OPAL, Physica B: Condensed Matter 385–386 Part 2 (2006) 1010-1012. https://doi.org/10.1016/j.physb.2006.05.322

[7] http://www.ncnr.nist.gov/instruments/darts/

[8] T. Gnaupel-Herold, P.C. Brand and H.J. Prask, Calculation of Single-Crystal Elastic Constants for Cubic Crystal Symmetry from Powder Diffraction Data, J. Appl. Crystallogr. 31 (1998) 929-935. https://doi.org/10.1107/S002188989800898X

[9] M. Law, O. Kirstein and V. Luzin, An assessment of the effect of cutting welded samples on residual stress measurements by chill modelling, J. Strain Anal. Eng. Des. 45 (2010) 567-573. https://doi.org/10.1177/030932471004500807

[10] Y.C. Tsui and T.W. Clyne, An analytical model for predicting residual stresses in progressively deposited coatings Part 1: Planar geometry, Thin Solid Films 306 (1997) 23-33. https://doi.org/10.1016/S0040-6090(97)00199-5

MECA SENS 2017   Materials Research Forum LLC
Materials Research Proceedings **4** (2018) 135-140   doi: http://dx.doi.org/10.21741/9781945291678-21

Performance Effects of Laser Deposited Ti-Al-Sn Coating on ASTM A29 Steel

O.S. Fatoba[1,a*], S.A. Akinlabi[2,b] and E.T. Akinlabi[1,c]

[1]Department of Mechanical Engineering Science, Faculty of Engineering and the Built Environment, University of Johannesburg, South Africa

[2]Department of Mechanical and Industrial Engineering Technology, Faculty of Engineering and the Built Environment, University of Johannesburg, South Africa

[a]drfatobasameni@gmail.com, [b]stephenakinlabi@gmail.com, [c]etakinlabi@uj.ac.za

Keywords: ASTM A29 Steel, Hardness, Wear, Ti-Al-Sn Coating, Intermetallics

Abstract. The conventional surface modification and coating cannot always fulfil the performance of material surface under extreme corrosion and wear environments. Corrosion and wear phenomenon lead to the gradual deterioration of components in industrial plants that can result in loss of plant efficiency, and even total shutdown with aggravated damage in industries. Hence, surface modification by incorporating chemical barrier coatings can be beneficial to this extent we report on investigation aimed at enhancing the surface properties of ASTM A29 steel by incorporating Ti-Al-Sn coatings deposited by laser deposition technique. For this purpose, a 3-kW continuous wave ytterbium laser system attached to a KUKA robot which controls the movement during the alloying process was utilized to deposit coatings with stoichiometry Ti-30Al-20Sn and Ti-20Al-20Sn. The alloyed surfaces were investigated in terms of its hardness and wear behaviour as function of the laser processing conditions. Hardness measurements were done using a vickers micro-hardness tester model FM700. Wear tests were performed on prepared ASTM A29 steel substrate deposited sample using the reciprocating tribometer (CERT UMT-2) under dry reciprocating conditions with continual recording of the dynamic coefficient of friction (COF) values. The microstructures of the coated and uncoated samples were characterized by optical and scanning electron microscopy. In addition, X-ray diffraction was used to identify the phase's contents. The optimum performances were obtained for an alloy composition of Ti-20Al-20Sn, at laser power of 750 W and coating speed of 0.8 m/min. Its performance enhancement compared to the unprotected substrate comprised a significant increase in hardness from 115 to 509 HV and reduced wear volume loss from 0.717 to 0.053 mm^3. The enhanced performance is attributed to the formation of the intermetallic phases Ti_6Sn_5, $AlSn_2Ti_5$, Ti_3Al, and TiAl.

Introduction

Al–Sn based alloys are widely used as sliding bearing materials in automobile and shipbuilding industry due to their good compact ability, wear resistance, thermal conductivity, and sliding properties [1]. The alloys also excel in high temperature stability. In these alloy systems, tin is a necessary soft phase in the aluminium matrix. Due to its low modulus, low strength and the excellent anti-welding characteristics with iron, tin phase in Al–Sn bearing materials can provide suitable frictional properties and shear surface during sliding [1]. The development of uniform microstructures with improved performance has been necessitated by the growing importance of Al-Sn based alloys as materials for engineering applications [2]. However, the processing of these alloys by conventional liquid metallurgy routes results in coarse grain microstructure with large degree of segregation of alloying elements [3]. Nevertheless, literature on hardness and

MECA SENS 2017   Materials Research Forum LLC
Materials Research Proceedings 4 (2018) 135-140   doi: http://dx.doi.org/10.21741/9781945291678-21

wear resistance performance of Ti-Al-Sn alloy coatings on ASTM A29 steel by laser surface alloying (LSA) technique are very scarce. LSA can rapidly provide a thick and crack-free layer in all instances with metallurgical bonds at the interface between the alloyed layer and the substrate. Powders surfaced on new or worn working surfaces of components by LSA provides specific properties such as high abrasive wear resistance, erosion resistance, corrosion resistance, heat resistance and combinations of these properties. Consequently, improvements in machinery performance and safety in aerospace, automotive, can be realized by the method [4]. The present study investigates the effect of laser processing parameters on the hardness and wear resistance performance of Ti-Al-Sn coatings on ASTM A29 steel.

Experimental details
Materials Specifications and Sample Preparation Method. The substrate material used in the present investigation was ASTM A29 steel. The substrate was cut, and machined into dimensions 100 x 100 x 5 mm^3. Prior to laser treatment, the substrates (ASTM A29 steel) were sandblasted, washed, rinsed in water, cleaned with acetone and dried in hot air before exposure to laser beam to minimize reflection of radiation during laser processing and enhance the absorption of the laser beam radiation. Ti (99.9 % purity), Al (99.9 % purity) and Sn (99.9 %) reinforcement metallic powders were mixed in 60:20:20(A1), 60:20:20(A2), 50:30:20(B1), 50:30:20(B2) ratio, respectively, in a shaker mixer (Turbular T2F; Glenn Mills, Inc.) for 12 hours at a speed of 49 rpm to obtain homogeneous mixture. The particle shape of the powder used was spherical with 50-105 μm particle sizes.

Wear tests were performed on the deposited sample at room temperature using the reciprocating tribometer (CERT UMT-2; Bruker Nano Inc., Campell, CA) under dry reciprocating conditions with continual recording of the dynamic coefficient of friction values. The normal load applied on the samples was 25 N at a frequency of 5 Hz and 2 mm stroke length using tungsten carbide (WC) counter material. Laser surface alloying was performed using a 3-kW continuous wave (CW) Ytterbium Laser System (YLS) controlled by a KUKA robot which controls the movement of the nozzle head and emitting a Gaussian beam at 1064 nm. The nozzle was fixed at 3 mm from the steel substrate. The admixed powders were fed coaxially by employing a commercial powder feeder instrument equipped with a flow balance to control the powder feed rate. The metallic powder was fed through the off-axes nozzle fitted onto the Ytterbium fibre laser and it was injected simultaneously into a melt pool formed during scanning of the ASTM A29 steel by the laser beam. Argon gas flowing at a rate of 2.5 L/min was used as a shielding gas to prevent oxidation of the sample during laser surface alloying. Overlapping tracks were obtained by overlapping of melt tracks at 70 %. To determine the best processing parameters, optimization tests were performed with the laser power of 750 to 900 W and scanning speed varied from 0.6 to 0.8 m/min. The final selection criteria during optimization tests was based on surface having homogeneous layer free of porosity and cracks determined from SEM analysis. The optimum laser parameters used was 900 W power, a beam diameter of 3 mm, gas flow rate of 2.5 L/min, powder flow rate of 2.0 g/min and scanning speeds of 0.6 m/min and 0.8 m/min respectively.

According to Qu and Truhan [5], the wear depth Zw and wear volume of the flat specimen can be calculated with the following equations:

$$V_W = L_S \left[R_S^{\ 2} \sin^{-1}(\frac{W}{2R_S}) - \frac{W}{2}(R_S - Z_W) \right] + \frac{\pi}{3} Z_W^{\ 2}(3R_S - Z_W) \tag{1}$$

$$Z_W = R_S - \sqrt{R_S^2 - \frac{W^2}{4}} \tag{2}$$

Where Z_W = Wear depth; R_S = Radius of spherical surface at both ends; W = Width of the wear scar

V_W = Wear Volume; L_S = Stroke length

Results and discussion

Morphological and Phase Analyses of Ti-Al-Sn Ternary Coatings. Intermetallic phases Ti, Al_5Ti_2, Al_3Ti, and Sn_5Ti_6 are common to alloyed samples at laser speed of 0.6 and 0.8 m/min. More peaks with smaller interspacing distance of order 2.75 to 1.34 Å were visible in the XRD spectrum showed in Fig. 1. Furthermore, there is good evidence of Al, Sn and Ti metal in the XRD spectrum. The presence of Al, Sn and Ti after laser alloying process formed three major hard phases of aluminium-titanium (AlTi$_2$, Al_5Ti_2, Al_3Ti, $AlTi_3$, Al_2Ti, Al_2Ti), tin-titanium (Sn_5Ti_6), and aluminium-tin-titanium ($AlSn_2Ti_5$). The phases formed showed good interfacial bonding and evidence of reactions occurring between Al and Ti, Al, Sn and Ti, Sn and Ti. The reaction between elemental powders of Ti and Al led to the formation of TiAl$_3$, Ti$_3$Al, and TiAl according to the binary phase diagram of Ti-Al. The formation of titanium-aluminides intermetallics took place through an exothermic reaction between solid titanium and liquid aluminium [6]. On the other hand, TiAl$_2$ and Ti$_2$Al$_5$ would require TiAl as an intermediate product for their formation [6].

Grain refinement effect of titanium which plays a vital function in influencing the critical properties of aluminium products have been studied by previous researchers. It enhances plasticity and tensile intensities and reduces the tendency of porosity and hot tearing [7]. This is due to the peritectic reaction occurrence at the end of aluminium rich in aluminium-titanium phase diagram [8]. It was clearly observed from the XRD spectrum of laser alloyed samples in Fig. 1 the presence of aluminium-titanium phases in five different forms such as AlTi$_2$, Al_5Ti_2, Al_3Ti, $AlTi_3$, and Al_2Ti. The aluminium-titanium phases increase as the laser speed increases from 0.6 to 0.8 m/min. It indicates that as the speed decreases, there is good possibility for Al to react with Ti. In addendum, titanium-aluminides such as Ti$_3$Al and TiAl exhibit significant potential to be a good alternative to existing conventional iron-aluminides, titanium alloys, and nickel super-alloys [9]. Fig. 1 shows intermetallic compounds with evidence of aluminium-titanium (AlTi$_2$, Al_5Ti_2, Al_3Ti, $AlTi_3$, Al_2Ti), tin-titanium (Sn_5Ti_6), and aluminium-tin-titanium ($AlSn_2Ti_5$). It can also be observed that as the laser speed increases from 0.6 to 0.8 m/min there are smaller grain sizes of the various phases formed. This could be attributed to the fact that increases in laser speed led to faster cooling of the melt pool which resulted into the fine grain sizes as shown in Fig. 2.

Moreover, according to Akinlabi and Akinlabi [10], increase in number of scan changes to type of heat treatment and produces strain hardening in material causing the grain sizes to be reduced as laser scans increases. In addendum, increase of the scanning speed results in finer microstructure due to the larger cooling rate during solidification as reported by Gong *et al.* [11]. Fig. 2 shows a microstructure without pores. It is well known that porosity results from gas bubbles trapped in the melt pool when the front wall solidifies [12] and this is influenced by laser scanning speed.

Materials Research Proceedings **4** (2018) 135-140   doi: http://dx.doi.org/10.21741/9781945291678-21

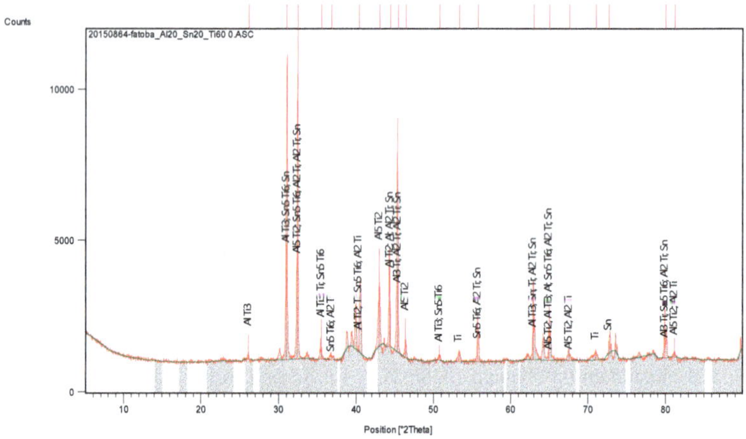

Fig. 1: XRD Spectrum of Ti-20Al-20Sn-0.8 Ternary Coating.

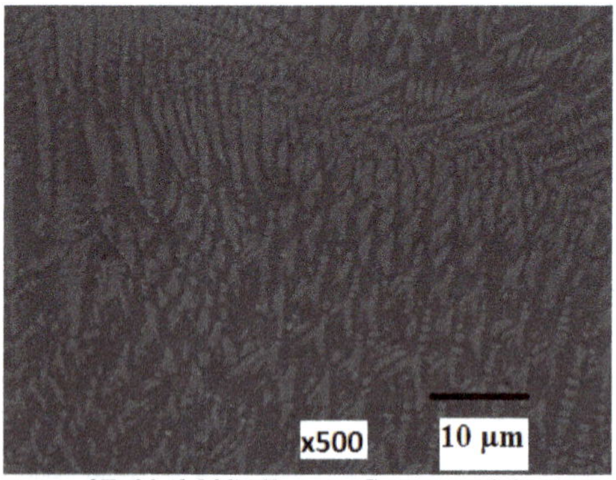

Fig. 2: SEM Images of Ti-20Al-20Sn Ternary Coating at 0.8 m/min scanning speed.

Microhardness Property of Ti-Al-Sn Ternary Coatings. The results showed that the laser alloying process enhances the hardness value of the substrate as shown in Fig. 3. Hardness values range between 115 to 509 HV. The hardness values of 115, 288, 376, 476, and 509 HV were obtained for substrate, Ti-30Al-20Sn-0.6, Ti-30Al-20Sn-0.8, Ti-20Al-20Sn-0.6, and Ti-20Al-20Sn-0.8 respectively. A raise of 75.84 and 81.51 % in hardness values above that of the substrate at Ti-20Al-20Sn-0.6 and Ti-20Al-20Sn-0.8 respectively. This increased hardness values are attributed to the hard phases of aluminium-titanium ($AlTi_2$, Al_5Ti_2, Al_3Ti, $AlTi_3$, Al_2Ti, Al_2Ti), tin-titanium (Sn_5Ti_6), and aluminium-tin-titanium ($AlSn_2Ti_5$) formed after the laser alloying process as evident by the XRD spectrum and SEM image shown in Figs. 1 and 2.

MECA SENS 2017   Materials Research Forum LLC
Materials Research Proceedings 4 (2018) 135-140   doi: http://dx.doi.org/10.21741/9781945291678-21

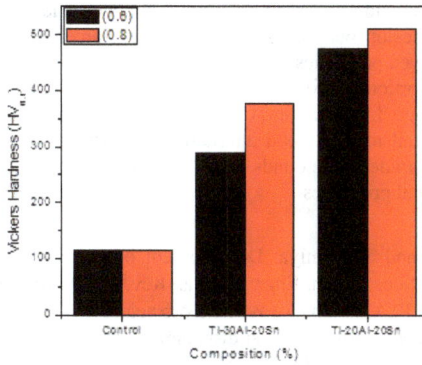

Fig. 3: Comparative Microhardness Chart of the Control and Al-Sn-Ti Ternary Coatings.

Wear Performance of Ti-Al-Sn Coatings. Comparisons of the experimental results show that the friction coefficient attained by ternary coatings was remarkable and range between 0.19 and 0.42 respectively as shown in Fig. 4. In general, the friction coefficient of coated samples indicated remarkable improvement in wear resistance performance compared to the control sample, with 35.38 % and 70.77 % reduction in coefficient of friction of Ti-20Al-20Sn-0.6 and Ti-20Al-20Sn-0.8 respectively. With Eq. 1, the wear volume losses of the control, and coated samples (Ti-20Al-20Sn-0.6, Ti-20Al-20Sn-0.8, Ti-30Al-20Sn-0.6, Ti-30Al-20Sn-0.8) were calculated as 0.72, 0.075, 0.053, 0.083, and 0.089 mm^3 respectively.

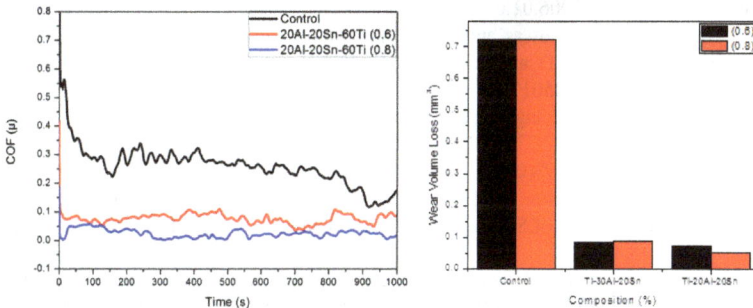

Fig. 4: Variation of the Friction Coefficient with Time and Wear Volume Loss for the Control and Al-Sn-Ti Ternary Coatings.

All the ternary coatings samples showed decrease in plastic deformation with Ti-20Al-20Sn-0.8 indicating an outstanding decrease in adhered layer and surface dislocation. The wear losses of ternary coated samples indicated remarkable improvement in wear resistance performance compared to the control sample with 90.38 % and 92.68 % reduction in wear volume losses of Ti-20Al-20Sn-0.6 and Ti-20Al-20Sn-0.8 respectively.

Conclusion

Well optimized process parameters and carefully chosen reinforcement materials fractions produced coatings with enhanced hardness and wear resistance properties. Crack formation was eliminated through optimization of laser processing parameters, leading to enhanced quality of the coatings, surface adhesion between substrate and reinforcement materials, microstructural evolution and thus improved properties.

The composition proportion of mixed powders has a great influence on the phase structure of the laser deposited coatings. In addendum, titanium-aluminides such as Al_3Ti, and $AlTi_3$ formed exhibit significant potential to be a good alternative to existing conventional iron-aluminides. Different titanium aluminide compounds such as $TiAl_3$, Ti_3Al and $TiAl$ also influence tribological and mechanical properties.

References

[1] T. Desaki, Y. Goto and S. Kamiya, Development of the Aluminium Alloy Bearing with Higher Wear Resistance, Soc. Autom. Engr. of Japan Rev. 21 (2000) 321-325.

[2] V. Bhattacharya and K. Chattopadhyay, Microstructure and wear behaviour of aluminium alloys containing embedded nanoscaled lead dispersoids. Acta Materialia. 52 (2004) 2293-2304. https://doi.org/10.1016/j.actamat.2004.01.020

[3] X.J. Ning, J.H. Kim, H.J. Kim, C.J. Li and C. Lee, Characteristics and heat treatment of cold-sprayed Al-Sn binary alloy coatings, Surf. Coat. Technol. 202 (2008) 1681. https://doi.org/10.1016/j.surfcoat.2007.07.026

[4] O.S. Fatoba E.T. Akinlabi and M.E. Makhatha. Effect of Process Parameters on the Microstructure, Hardness, and Wear Resistance Properties of Zn-Sn-Ti Coatings on AISI 1015 Steel: Laser Alloying Technique. Int. J. Surf. Sci. Eng. 11 (6) (2017) 489-511. https://doi.org/10.1504/IJSURFSE.2017.088969

[5] J. Qu and J.J. Truhan, An efficient method for accurately determining wear volumes of sliders with non-flat wear scars and compound curvatures, Wear 261 (2006) 848–855. https://doi.org/10.1016/j.wear.2006.01.009

[6] M. Sujata, S. Bhargava, S. Suwas and S. Sangal. On Kinetics of TiAl3 Formation during Reaction Synthesis from Solid Ti and Liquid Al, J. Mater. Sci. Lett. 20 (2001) 2207-2209. https://doi.org/10.1023/A:1017985017778

[7] X.F. Liu, Z.Q. Wang, Z.G. Zhang and X.F. Bian. The Relationship between Microstructures and Refining Performances of Al-Ti-C Master Alloys, Mater. Sci. Eng. A 332 (2002) 70-74. https://doi.org/10.1016/S0921-5093(01)01751-8

[8] R.O. Kaibyshev, I.A. Mazurina and D.A. Gromov, Mechanisms of Grain Refinement in Aluminum Alloys in the Process of Severe Plastic Deformation, Met. Sci. Heat Treat. 48 (2006) 57-62.

[9] R.L. Fleischer, D.M. Dimiduk and H.A. Lipsitt, Intermetallic Compounds for High Temperature Materials: Status and Potential, Annu. Rev. Mater. Sci. 19 (1989) 231-263. https://doi.org/10.1146/annurev.ms.19.080189.001311

[10] E.T. Akinlabi and S.A. Akinlabi, Effect of Heat Input on the Properties of Dissimilar Friction Stir Welds of Aluminium and Copper, Am. J. Mater. Sci. 2 (2012) 147-152. https://doi.org/10.5923/j.materials.20120205.03

[11] X. Gong, J. Lydon, K. Cooper, and K. Chou, Beam Speed Effects on Ti-6Al-4V Microstructures in Electron Beam Additive Manufacturing, J. Mater. Res. 29(17) (2014) 1951-1959. https://doi.org/10.1557/jmr.2014.125

[12] W. Meng, Z. Li, F. Lu, Y. Wu, J. Chen and S. Katayama, Porosity Formation Mechanism and its Prevention in Laser Lap Welding. J. Mater. Process. Technol. 214 (2014) 1658-1664. https://doi.org/10.1016/j.jmatprotec.2014.03.011

Techniques & Instruments

MECA SENS 2017
Materials Research Proceedings 4 (2018) 143-148

Materials Research Forum LLC
doi: http://dx.doi.org/10.21741/9781945291678-22

Alignment and Calibration Procedures of the Necsa Neutron Strain Scanner

D. Marais[1,a*], A.M. Venter[1,b] and V. Luzin[2,c]

[1]Research and Development Division, Necsa SOC Limited, Pretoria, South Africa

[2]Australian Centre for Neutron Scattering, ANSTO, Lucas Heights, NSW, Australia

[a]deon.marais@necsa.co.za, [b]andrew.venter@necsa.co.za, [c]vll@ansto.gov.au

Keywords: Neutron Diffraction Instrument, Strain Scanner, Alignment and Calibration

Abstract. The alignment and calibration of neutron strain instruments directly affect the reliability and accuracy of neutron stress experiments. Procedures for the purpose of instrument optimization and characterization were established at Necsa's angular dispersive neutron strain scanner but may also be applicable to other diffraction instruments. Attention is given to the following aspects: sample table center of rotation, horizontally focusable multi-wafer silicon single crystal monochromator, neutron area detector offset and the instrument gauge volume.

Introduction

The Materials Probe for Internal Strain Investigations (MPISI) neutron diffraction strain scanner [1] is situated at the SAFARI-1 research reactor of the South African Nuclear Energy Corporation (Necsa) SOC Limited and is depicted in Fig. 1. A commissioning phase followed after a recent major instrument upgrade to improve measurement accuracy and precision.

In order to maximize the available neutron flux at the sample position and reduce the error in the measured diffraction angle (and consequently in the d_{hkl} and calculated stress value), all components contributing to the neutron optical path of the instrument must be aligned to as high an accuracy as practically achievable. Complete instrument alignment is generally an iterative process as the system comprises of a number of interdependent components.

Center of rotation

The instrument reference point is measurable and is defined as the intersection of the sample table's center of rotation (CoR) axis with the diffraction plane. In order to determine the CoR, the linear x and y-axes (motors sx and sy) of the sample table should be adjusted such that an accurately machined calibration pin is not be displaced by more than 10 % of the applicable gauge volume dimension when

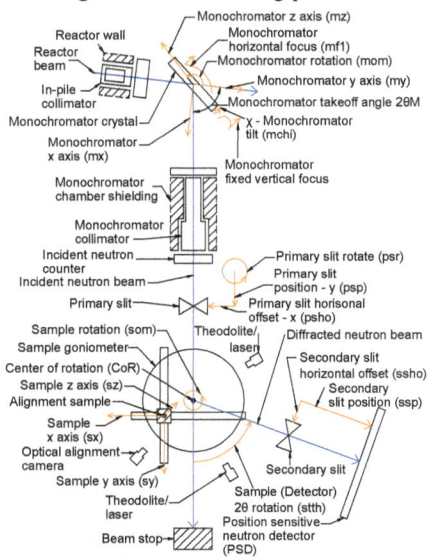

Fig. 1: Diagram of MPISI indicating selected motor names and components.

driving the sample table's omega rotation axis (som motor) [2]. The pin location can be measured either using a digital dial gauge [3] or a telecentric camera alignment system [4]. Communication modules for both the digital dial gauge and the camera system were created in the MPISI control system SICS [5] in order to automate alignment procedures.

After the axes of the measurement device are aligned with the axes of the sample positioning table, the x and y-axes offsets needed to move the alignment pin to the CoR must be determined. This can be done using an iterative procedure to solve a system of three equations [6], but this method is heavily dependent on starting the solution process with good initial values. In order to overcome this limitation, explicit equations (Eqs. 1 and 2) were developed to calculate the offsets by using four measurements taken 90° apart (Fig. 2).

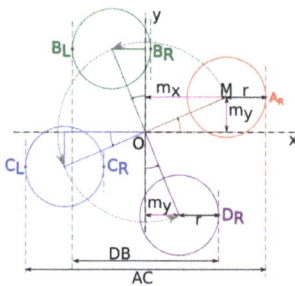

$$m_x = \frac{A_R - C_R}{2}. \tag{1}$$

$$m_y = \frac{D_R - B_R}{2}. \tag{2}$$

where:
$m_x = x$-axis offset
$m_y = y$-axis offset
$A_R = $ Gauge reading at position A (0°)
$B_R = $ Gauge reading at position B (90°)
$C_R = $ Gauge reading at position C (180°)
$D_R = $ Gauge reading at position D (270°)

Fig. 2: Depiction of the displacement of a misaligned pin about the CoR when rotated though 270° at 90° intervals.

Explicit equations (Eq. 3 and 4) can also be used to determine the offsets by using three measurements taken 45° apart (Fig. 3). This method is very convenient when a large sample omega rotation is not possible such as when a mounted sample is restricting movement.

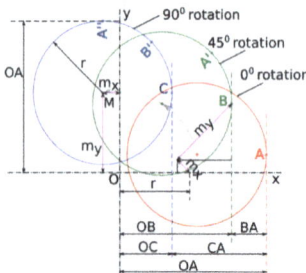

$$m_x = \frac{1}{\sqrt{2}-1}\left(BA - CA + \frac{CA}{\sqrt{2}}\right). \tag{3}$$

$$m_y = CA - \frac{1}{\sqrt{2}-1}\left(BA - CA + \frac{CA}{\sqrt{2}}\right). \tag{4}$$

where:
$m_x = x$-axis offset
$m_y = y$-axis offset
BA, $CA = $ Measured distance (x-projection) between positions B and A, and C and A respectively

Fig. 3: Depiction of the displacement of a misaligned pin about the CoR when rotated though 90° at 45° intervals.

Both methods were implemented on MPISI. The final pin position (relative) as measured with the digital dial gauge after using the camera alignment system is given in Fig. 4. When using the camera alignment system, the 3-measurement method should therefore not be used for gauge volume dimensions smaller than 0.5 mm.

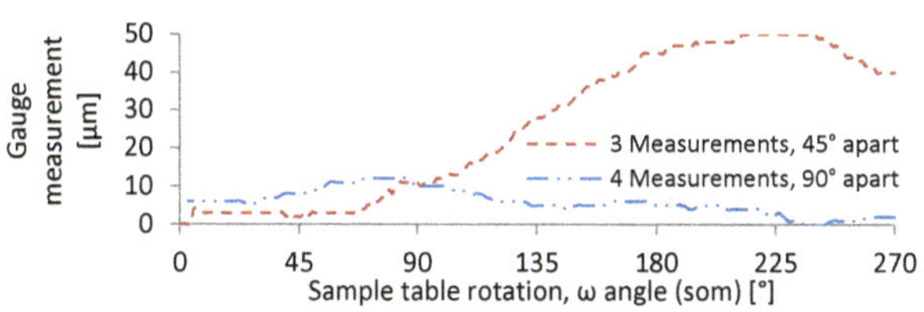

Fig. 4: Relative pin position when rotating the sample table after two different CoR alignment procedures.

Monochromator alignment

Ideally, the monochromator must be positioned such that the point where the monochromator assembly's center of rotation (CoR_M) and tilt rotation center intersects is located at the intersection of the reactor beam and the required primary neutron beam. An approximation of the direction of the primary beam can be visually determined by replacing the monochromator collimator with a transparent Perspex plug that has a small hole (~2 mm diameter) drilled through its center. A laser is then directed through the hole into the monochromator chamber and the x, y, and z offsets (corresponding to the mx, my and mz motors) of the assembly are adjusted until the laser spot is observed on the center of the monochromator face. With reference to Fig. 6, x and y are respectively tangential and parallel to the reactor beam with z out of the page. The *tilt* axis (mchi motor) enables the monochromator to lean forward and backward.

With the Si (110) plane of the monochromator cut parallel to its face, the planes that are diffracted from when rotating the crystal in the horizontal plane about [$\bar{1}10$] within the constraints of the available space inside the monochromator chamber, are depicted in the stereographic projection given in Fig. 5. Si (110) was chosen due to a fixed monochromator takeoff angle through the monochromator chamber and space restrictions on the beam-port floor. By using this cut plane, appropriate reflections can be reached to obtain close to 90° diffraction angles from typical engineering materials. Silicon has a cubic diamond crystal structure and therefore the angle φ between two planes ($h_1k_1l_1$) and ($h_2k_2l_2$) is given by Eq. 5 [7]. Using this equation, the angles between diffracting planes were calculated and are depicted in Fig. 6.

$$\varphi = \cos^{-1}\left(\frac{h_1h_2 + k_1k_2 + l_1l_2}{\sqrt{h_1^2 + k_1^2 + l_1^2}\sqrt{h_2^2 + k_2^2 + l_2^2}}\right). \tag{5}$$

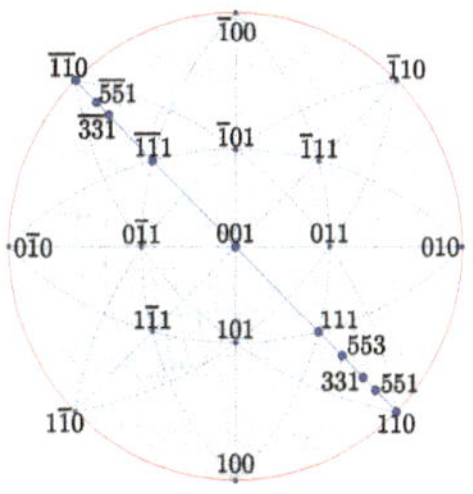

Fig. 5: Stereographic projection of a diamond cubic crystal structure showing poles relevant to MPISI's Si monochromator.

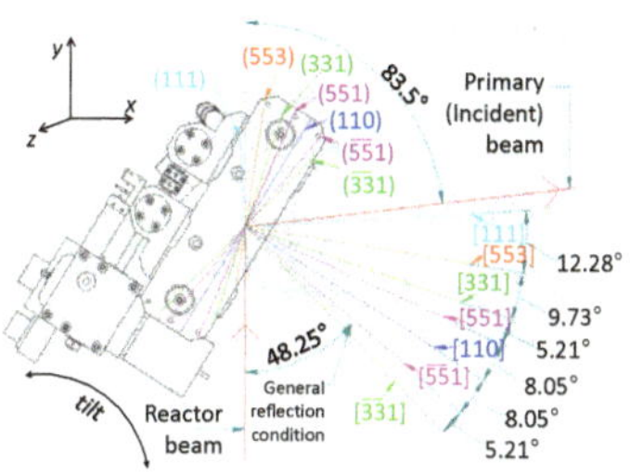

Fig. 6: Diagram of the MPISI Si crystal monochromator showing the orientation and direction of the crystal planes including the diffraction condition for a 83.5° take-off angle.

The monochromator rotation angle (mom motor) was step-scanned whilst recoding the neutron count rate provided by a neutron counter positioned in the primary beam. The resulting peaks (Fig. 7) were fit with Gaussian functions to determine the diffraction angles. The separation angles between peaks were then used to identify (index) the peaks. It should be noted that the structure factor for the (110) plane of Si is 0, therefore the diffraction observed at the associated angle is primarily from the (220) reflection.

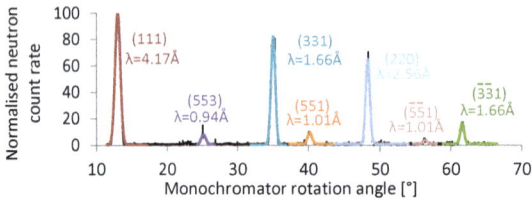

Fig. 7: Graph of the normalized neutron count rate at the monochromator exit port as a function of monochromator rotation angle. The primary wavelength for each peak is also given.

The (111) peak was selected to align the monochromator for maximum intensity. The monochromator mx, my and mchi motors were sequentially step-scanned over the achievable translation ranges whilst recording the neutron count rate on the beam monitor. After each scan, a Gaussian function was fit to the acquired data in order to determine the motor position rendering maximum count rate.

A well-aligned monochromator not only delivers a high intensity neutron beam but also minimizes the spread in wavelength to the sample position. There is however a trade-off between observed integrated intensity and spread in wavelength which can be optimized using the figure of merit (FOM) [8] given in Eq. 6.

$$FOM \propto \frac{Intensity}{FWHM^2}. \tag{6}$$

In order to optimize the FOM for the diffraction angle ($2\theta_S$) range commonly used at MPISI, a mild steel pin was positioned on the CoR and the monochromator curvature (motor mf1) was step-scanned from 0.04 m^{-1} to 0.61 m^{-1} in 574 steps whilst recording the diffraction pattern on the neutron detector. The monochromator was positioned on the Si (331) reflection. Fig. 8a and b respectively show the Fe (211) diffraction peak and the resulting FOM as a function of horizontal monochromator curvature. The optimum curvature was obtained from the FOM by fitting a Gaussian function to the data.

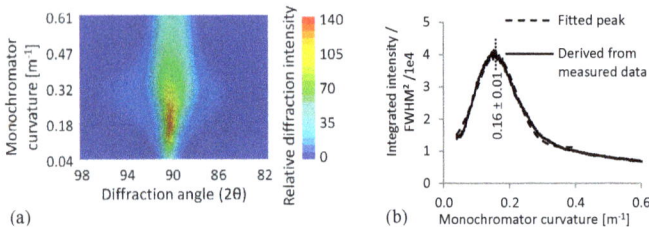

Fig. 8: Graphs of (a) the Fe (211) diffraction peak and (b) the resulting figure of merit, as a function of monochromator curvature.

The diffraction plane was aligned with the center of the detector by ensuring that the center of a low-angle diffraction cone coincides with the vertical center of the detector. To achieve this, a Mo (110) peak was measured at $43.8°$ $2\theta_S$ using neutrons with a wavelength of 1.659 Å. The observed diffraction cone was then subdivided into 30 vertical sections (Fig. 9a) and each section

MECA SENS 2017   Materials Research Forum LLC
Materials Research Proceedings **4** (2018) 143-148   doi: http://dx.doi.org/10.21741/9781945291678-22

integrated to produce a diffraction peak corresponding to a *y-detector* position. A Gaussian function was fit to each peak to determine the associated diffraction angles. These diffraction angles (as a function of *y-detector* position) was again fit with a Gaussian function to determine the center of the diffraction cone with respect to the detector position as is shown in Fig. 9.b.

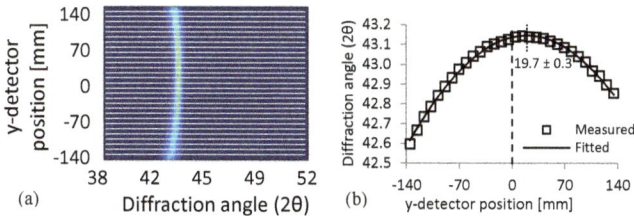

Fig. 9: (a) Diffraction cone of Mo (110) plane intercepting the detector and (b) a Gaussian fit over the diffraction cone showing the center position.

By changing the monochromator tilt angle, the cone direction can be adjusted as the vertical incident angle onto the sample is changed. When the tilt angle is changed, the monochromator vertical offset (*z*-axis position) should also be adjusted to ensure that the beam is not restricted by the monochromator collimator. An iterative process therefore follows for the *tilt* and *z*-axis alignment.

Detector alignment
The first step in aligning the detector is to establish the relationship between the detector bins (chosen as 421×421) and the detector active area. We know from the detector specification that the active area is 280×280 mm^2, therefore each bin represents an area of 0.67×0.67 mm^2.

In order to calculate the sample-to-detector distance, it is necessary to determine the total detector angular span. A mild steel calibration pin was mounted at the CoR and the detector positioned with the detector $2\theta_S$ angle (stth motor) at 94° in order for the Fe (211) diffraction peak to be observed at the left edge of the detector. The detector was step-scanned in 4 x 2° steps to $2\theta_S = 86°$ and each peak were fit with a Gaussian function to determine the *x-detector* position of the peak center. The linear relationship between the observed peak positions as a function of $2\theta_S$ angle was used to calculate the total detector angular span as 13.9°. The sample to detector distance was established as 1148.5 mm by using the angular span and detector active length.

Al$_2$O$_3$ NIST standard powder was positioned on the sample table and the detector $2\theta_S$ angle step-scanned in 6 steps from 35° to 100°. After stitching the acquired frames, Rietveld refinement was applied to the resulting diffraction pattern and the detector offset and neutron wavelength was determined as -1.78° and 1.646 Å respectively. Using this wavelength, the monochromator exit port angle ($2\theta_M$) was calculated as 82.7°, instead of the estimated 83.5°.

Instrument gauge volume
The vertical center of the beam was determined by *z*-step scanning (using the sz motor) a horizontal positioned mild steel bar through the primary beam with the apertures removed. The neutron counts from the Fe (211) diffraction peak were integrated and an analytical solution was fit to the measured data as a function of sz motor position. The center position was thereby resolved to within 0.2 mm. The height of the lasers, theodolites and apertures were then adjusted to coincide with the bar center.

The two aperture horizontal positions were aligned by step scanning the slit horizontal offset axes (psho and ssho motors) respectively whilst recording the Fe (211) peak from a calibration cell positioned on the CoR. This measurement was done with each aperture at two different distances from the sample to calculate the mounting arm rotation required to ensure that the aperture always traverse parallel to the beam.

MPISI's beam divergence was subsequently measured by scattering from a mild steel bar using a nominal gauge volume of $2 \times 20 \times 2$ mm^3. Vertical and horizontal divergence were calculated by step-scanning the z and y sample table axes respectively as a function of primary slit position and fitting appropriate entry curve analytical functions to the integrated diffraction peak intensities. A horizontal divergence of 0.166° and vertical divergence of 0.458° were established.

Summary

Complete diffraction instrument alignment requires a systematic approach for alignment of individual components to establish a fixed instrument reference point and to optimize and characterize the incident (primary) and diffracted neutron beams. A number of these alignment steps are only performed during instrument commissioning eliminating the need for automation. There is however a number of routinely used procedures which can be automated in order to reduce human error and reduce the time needed to perform the alignment steps. The alignment and calibration procedures implemented at MPISI may also apply to other neutron diffraction instruments, depending on their specific configurations.

References

[1] A.M. Venter, P.R. van Heerden, D. Marais and J.C. Raaths, MPISI: The neutron strain scanner Materials Probe for Internal Strain Investigations at the SAFARI-1 research reactor, Physica B, In press. https://doi.org/10.1016/j.physb.2017.12.011

[2] G.A. Webster and R.W. Wimpory, (Eds.), Polycrystalline materials determination of residual stresses by neutron diffraction. Geneva: ISO, Technology Trends Assessment. (ISO/TTA3), 2002, pp. 29-37.

[3] Sylvac. Instructions for use – Dial Gauge S229. www.sylvac.ch Date of access: 14 May 2013 (2005)

[4] S. Flemming, Development of a distributed client server of system for the fully automatic instrument adjustment of a neutron diffractometer for residual stress analysis, Technische Fachhochschule Berlin, Berlin, 2012.

[5] H. Heer, M. Könnecke and D. Maden, The SINQ instrument control software system, Physica B 241-243 (1998) 124-126. https://doi.org/10.1016/S0921-4526(97)00528-0

[6] P.C. Brand and H.J. Prask, New methods for the alignment of instrumentation for residual-stress measurements by means of neutron diffraction, J. Appl. Cryst. 27 (1994) 164-176. https://doi.org/10.1107/S0021889893007605

[7] A. Kelly and K.M. Knowles, Crystallography and Crystal Defects. 2nd ed. West Sussex, UK: John Wiley & Sons, Ltd. 2012. https://doi.org/10.1002/9781119961468

[8] M.W. Johnson and M.R. Daymond, An optimum design for a time-of-flight neutron diffractometer for measuring engineering stresses, J. Appl. Cryst. 35 (2001) 49-57. https://doi.org/10.1107/S002188980101891X

MECA SENS 2017
Materials Research Proceedings **4** (2018) 149-154

Materials Research Forum LLC
doi: http://dx.doi.org/10.21741/9781945291678-23

Neutron Transmission Strain Measurements on IMAT: Residual Strain Mapping in an AlSiC$_p$ Metal Matrix Composite

R.S. Ramadhan[1,a*], W. Kockelmann[2,b], A.S. Tremsin[3,c] and M.E. Fitzpatrick[1,d]

[1]Centre for Manufacturing & Materials Engineering Coventry University, Coventry, CV1 2JH, UK

[2]STFC, Rutherford Appleton Laboratory, ISIS Facility, Chilton, OX11 0QX, UK

[3]Space Science Laboratory, University of California at Berkeley, CA 94720 Berkeley, USA

[a]ramadhar@uni.coventry.ac.uk, [b]winfried.kockelmann@stfc.ac.uk, [c]ast@ssl.bekeley.edu, [d]ab6856@coventry.ac.uk

Keywords: Neutron Transmission, Residual Strain, Strain Mapping, Metal Matrix Composite

Abstract. Neutron transmission strain measurements were performed for the first time at the IMAT beamline, ISIS, UK, in order to demonstrate the capability and measure the accuracy of the new instrument. A novel Bragg edge strain analysis technique based on cross correlation is introduced as an alternative to a Bragg edge fitting technique. Neutron transmission measurements were performed on an AlSiC$_p$ metal matrix composite, and the result is compared with the neutron diffraction result on an identical sample, showing good agreement between the two. The advantage of using the proposed cross correlation Bragg edge strain analysis technique over edge fitting is discussed.

Introduction

This paper reports the first attempt of neutron transmission strain analysis at IMAT (Imaging and Material Science), a new, combined imaging and diffraction instrument at the ISIS neutron spallation source, UK [1]. An AlSiCp metal matrix composite (MMC) was chosen as the sample of interest. The sample has a well-characterized residual strain profile, since a similar MMC sample has been previously measured using neutron diffraction [2], and therefore is an ideal test case to provide the level of accuracy of strain measurements on IMAT.

This paper also introduces a new Bragg edge analysis technique based on cross correlation to extract strain information from recorded neutron transmission signals. Until present day, the strain analysis is performed by determining the position of the Bragg edges using a non-linear fitting function. There are several different fitting functions available to describe Bragg edge, among them are functions discussed by Santisteban *et al.* [3] and Tremsin *et al.* [4]. Using the proposed cross correlation technique, the difference of position between any two Bragg edges (i.e. Bragg edge of sample and the reference) can be measured without the need of fitting the Bragg edge itself. The cross correlation technique could be beneficial to measure strain in the presence of texture, where a Bragg edge shape produced by the sample is distorted and difficult to describe with an analytical function.

Experimental

Materials and Specimens. The AlSiC$_p$ MMC sample is composed of Al 2124 matrix and pure silicon carbide particulate. The composite plate was made by evenly mixing 45 μm aluminum alloy powder and 3 μm silicon carbide powder, at 20 % weight fraction, then followed by complicated process including hot isostatic pressing, forging, rolling, cold water quenching and prolonged aging. Such heat treatment is expected to have introduced a parabolic residual strain variation through the thickness of the plate. The dimension of the sample is 35 mm × 35 mm ×

15 mm. The sample used in this study is a section taken from the same MMC plate used in the study by Fitzpatrick *et al.* [2]. Powders of the aluminum matrix and silicon-carbide reinforcement were used as stress-free reference samples.

Neutron transmission measurement. The MMC sample was studied using neutron transmission at IMAT, with a setup as illustrated in Fig. 1a. The incident "white neutron beam", with a wavelength energy spectrum ranging from $0.7 - 6.6$ Å, impinged onto the sample which was placed a few mm away from the microchannel plates (MCP) detector. The MCP detector consist of a ^{10}B-doped thin plate as neutron-electron convertor followed by a stack of thin plates with micro-pores as electron amplifier, and an array of 2×2 Timepix readout chips. Each chip has 55 μm × 55 μm pixels, and the detector provides 512×512 pixels in total, and a 28×28 mm^2 field of view. With the setup shown in Fig. 1a, the measured strain direction is parallel with the *x*-axis and averaged through the thickness of the sample along the transmission path. The coordinate system is shown in Fig. 1a, and each measurement was performed for 5 hours.

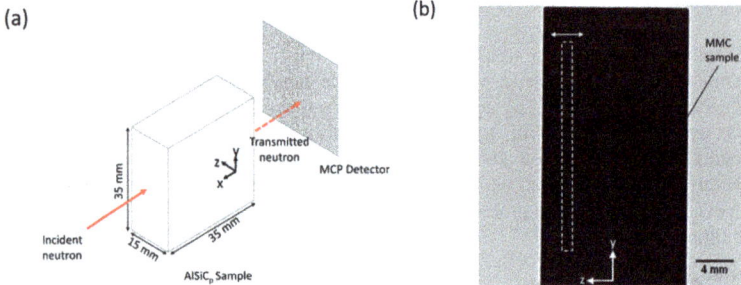

Fig. 1: (a) Setup of neutron transmission measurement; (b) Radiographic image of MMC sample, white dashed line showing averaging area for spatially-resolved strain measurement.

Bragg edge analysis for strain measurements
A Bragg edge, which signifies a sudden increase in neutron transmission, is formed as a consequence of backscattering phenomena [5]. The recorded transmission spectrum of AlSiC$_p$ MMC is shown in Fig. 2a, showing Bragg edges from both the aluminum matrix and the silicon-carbide reinforcement. In this work, two different analysis techniques were used to extract strain information from the Bragg edge signal. The first technique was to determine the strain using the shift in position between single Bragg edge from the sample and stress-free reference. The determination of the Bragg edge position was performed by fitting the measured data using two non-linear fitting functions described by Santisteban *et al.* [3] and Tremsin *et al.* [4], which will be called the Santisteban and Tremsin function, respectively, in the rest of this paper. An example of fitting results using the two fitting functions is shown in Fig. 2b.

The second technique to determine strain is based on applying a cross correlation algorithm between the sample Bragg edge and the stress-free reference Bragg edge data. Cross correlation is a mathematical function to measure similarity between two signals as function of the shift of one relative to the other. The analysis was carried out by initially taking the first derivative of both the sample and reference {111} Bragg edge signal. Then, cross correlation was performed on the two derivatives using Eq. (1):

$$y(m) = \sum_{n=0}^{M-1} f(n)g(n-m) = ifft(FG^*) \tag{1}$$

MECA SENS 2017   Materials Research Forum LLC
Materials Research Proceedings **4** (2018) 149-154   doi: http://dx.doi.org/10.21741/9781945291678-23

where f(n) and g(n) are the two Bragg edge derivatives, F is the Fourier transform of f(n), G is the Fourier transform of g(n), * means complex conjugation, ifft stands for inverse fast Fourier transform and y(m) is the correlation output. A Voigt function then is used to fit the peak-shaped correlation curve. The peak of the curve provides the information of position where the two Bragg edges are most similar, and thus inform on the shift (i.e. differences in position) between the edges. The shift is then used to calculate the strain.

Spatially resolved strain data were extracted by fitting the Bragg edges averaged over a region with 1 mm × 20 mm dimension, scanning along the z-axis, as shown in radiographic image of the sample, Fig. 1b. The selection area is elongated in y-axis direction in order to increase the statistics of the Bragg edge, assuming that there is not much strain variation as function of position along the y-axis. Strain mapping was performed by fitting the Bragg edge for all pixels on the dataset, using Tremsin function. Before fitting, neutron counts from several neighboring pixels over 2.2 mm × 2.2 mm area were combined into one spectrum, to improve the statistics [6], and maps were produced with a step size of 55 m using the 'running average' of this box.

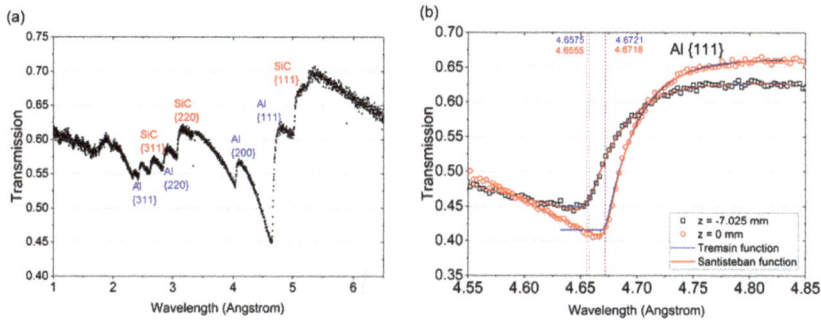

Fig. 2: (a) Transmission spectra of $AlSiC_p$ MMC sample; (b) Fitting of Al {111} Bragg edges for two different regions of interest using Santisteban and Tremsin functions.

Results

Residual strain values for the aluminum matrix of the $AlSiC_p$ MMC are shown in Fig. 3a. A parabolic strain profile through the thickness of the sample can be clearly seen where compressive strain is apparent on both surfaces of the plate, and with tensile strains in the middle. This is expected due to uneven material shrinkage caused by rapid cooling of the quenching process. The Tremsin function consistently gave slightly higher strain values compared to the Santisteban function, Fig. 3. Figure 3b show residual strain values for the silicon-carbide reinforcement component. It can be seen that the scatter of the data from silicon-carbide is higher than it is for aluminum. This is due to both lower levels of strain and a smaller volume fraction of silicon-carbide in the MMC sample that contributes to the Bragg edge signal. Although the measurement result is more scattered, a second-order polynomial fit of the strain measurement shows that the silicon-carbide component also has a parabolic-shaped strain profile, Fig. 3b.

Bragg edge strain analysis based on cross correlation was performed on the same dataset. The result was compared to strain analyzed with the Santisteban function. The strain measured by neutron transmission in this study was then compared with previous literature values measured by neutron diffraction on the similar MMC sample [2]. The comparison is shown in Fig. 4. In general, the neutron transmission shows very good agreement with the neutron diffraction result. The parabolic strain profile in the aluminum matrix measured by neutron diffraction is replicated almost with the exact magnitude and trend from neutron transmission, Fig. 4a. The strain analyzed by the cross correlation technique shows better agreement with neutron diffraction results. Cross correlation curves also have smaller errors compared to the non-linear fitting function results, Fig. 4a. Similarly, good agreement can be found between neutron diffraction and neutron transmission results for the silicon-carbide reinforcement phase, Fig. 4b. Despite the scatter of the data, neutron diffraction and neutron transmission strain values fall within the same range and follow the same trend.

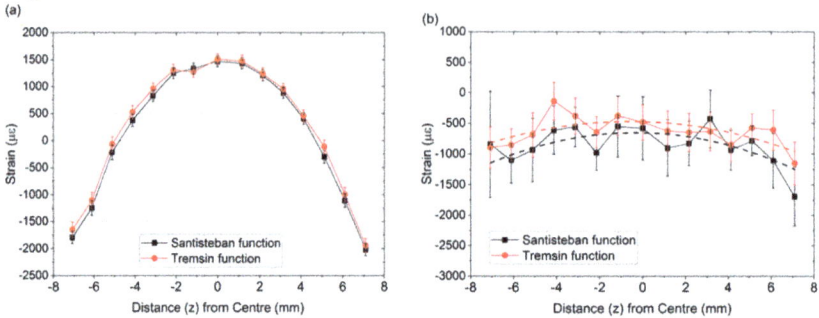

Fig. 3: x-axis residual strain as function of position along z-axis for (a) aluminum matrix and (b) silicon carbide reinforcement component of the $AlSiC_p$ MMC sample.

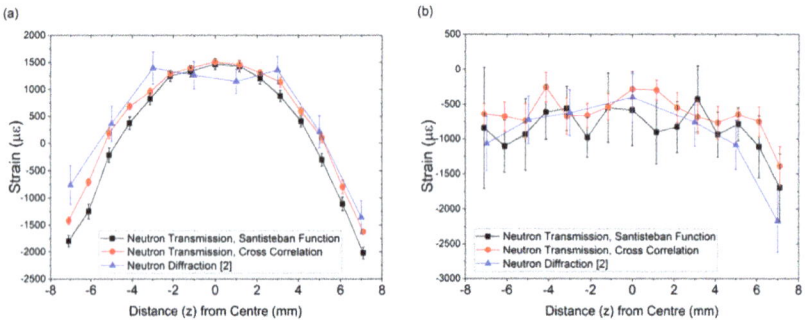

Fig. 4: Comparison between x-axis residual strain profiles obtained from neutron transmission using the non-linear fitting function, neutron transmission using cross correlation analysis, and neutron diffraction [2] for (a) aluminum and (b) silicon carbide component of MMC sample.

Figure 5 shows the reconstructed 2D residual strain maps within the MMC sample. A gradient of strain within the aluminum matrix through the thickness of the plate (z-axis direction) can be clearly observed, where compressive strain is apparent on both surfaces accompanied by tensile strain in the middle of the plate, Fig. 5a. Meanwhile, there is no significant strain variation along the y-axis direction, Fig. 5a. Figure 5b shows the residual strain map of the silicon-carbide component. The strain gradient through the thickness of the plate is less apparent due to higher

MECA SENS 2017   Materials Research Forum LLC
Materials Research Proceedings **4** (2018) 149-154   doi: http://dx.doi.org/10.21741/9781945291678-23

stiffness of silicon-carbide, although regions of tensile strain still can be found in the middle of the plate. Following the data analysis for strain mapping described in [6], the achieved spatial resolution is in range of a few hundred microns.

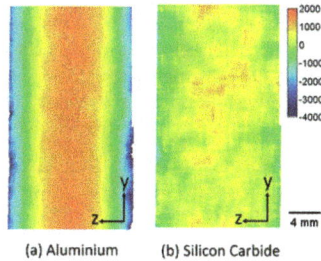

(a) Aluminium   (b) Silicon Carbide

Fig. 5: Residual strain map of AlSiC$_p$ MMC sample measured using neutron transmission.

Discussion

Fitting Bragg edges. Having more parameters to model the data, the Santisteban function can fit the Bragg edge better compared to the Tremsin function, especially for the data on the left hand side of the edge which shows negative exponential function, Fig. 2b. This leads to discrepancies within 190 μstrain between the fitting results, Fig. 3a. The slope of the exponential function on the left hand side of the Bragg edge is often influenced by preferred orientation present in the sample. As many engineering samples have some degree of preferred orientation, the Santisteban function is more robust and should be used for Bragg edge analysis.

Cross correlation. As mentioned above, the shape of a Bragg edge can be affected by the preferred orientation present in the sample. In the case where this texture effect is very strong, the Bragg edge shape is difficult to be modelled by an analytical function. The cross correlation technique proposed in this study eliminates the need of fitting the individual Bragg edges, and therefore theoretically allows strain analysis from Bragg edges of any shape. The early implementation of cross correlation in this study shows that this analysis technique performs as good as non-linear fitting functions. Bragg edge strain analysis using cross correlation and non-linear analytical functions yields comparable trends, with average differences in values less than 400 μstrain, Fig. 4. Moreover, especially for aluminum matrix data, cross correlation results shows better agreement with neutron diffraction, Fig. 4a. This might indicate that cross correlation measured strain more accurately in the presence of texture in the sample. The current limitation of this technique is the inability to analyse data with high noise. Using Eq. 1, cross correlation works better for peak-shaped data, and thus creating the need of performing the technique on the first derivative to the Bragg edge spectra. Consequently, this differentiation process amplifies the noise of the data. Therefore, at this point, the cross correlation works best for analysing statistically-good Bragg edge data.

Strain measurements on IMAT. The agreement between neutron transmission and neutron diffraction results shows the accuracy of IMAT instrument for strain measurement. The difference in values between IMAT results and neutron diffraction results for the sample studied is within ~500 μstrain using the non-linear fitting function technique and ~250 μstrain using the cross correlation technique, Fig. 4. Some of the difference is due to the fact that the neutron diffraction result was taken from a different cut of an identical plate, hence adding to the uncertainty. Also, the neutron transmission averaged the strain variation through the thickness of the sample, while neutron diffraction measured at a specific locations within the sample

MECA SENS 2017   Materials Research Forum LLC
Materials Research Proceedings 4 (2018) 149-154   doi: http://dx.doi.org/10.21741/9781945291678-23

thickness. Further studies will be conducted to investigate the accuracy and the resolution of the strain that can be measured at IMAT, whether using a round-robin reference sample [7], or measuring exactly the same sample using both neutron transmission and neutron diffraction. This study also demonstrated the capability of strain mapping on IMAT using the MCP detector. 2D strain maps can provide better understanding of local strain gradients within the sample. In this case it proves the initial assumption that there is not much strain variation along the *y*-axis direction, Fig. 5. Considering the sub-mm spatial resolution that was achieved, IMAT is able to produce the strain map in a relatively short time and in a simple experimental setup.

Conclusion

The capability and accuracy of the IMAT instrument for neutron transmission strain measurements has been demonstrated by the good agreement between neutron transmission and neutron diffraction results on two identical AlSiC$_p$ metal-matrix composite samples. A novel Bragg edge strain analysis based on cross correlation has been introduced, showing agreement within ~250 μstrain with neutron diffraction results. Comparatively, using a commonly used non-linear fitting function, neutron transmission shows agreement within ~500 μstrain with neutron diffraction. Accordingly, the strain mapping capability of IMAT has been demonstrated, and will be advantageous to study samples with more complicated residual strain distributions.

References

[1] T. Minniti, W. Kockelmann, G. Burca, J.F. Kelleher, S. Kabra, S.Y. Zhang, D.E. Pooley, E.M. Schooneveld, Q. Mutamba, J. Sykora, N.J. Rhodes, F.M. Pouzols, J.B. Nightingale, F. Aliotta, L.M. Bonaccorsi, R. Ponterio, G. Salvato, S. Trusso, C. Vasi, A.S. Tremsin and G Gorini, Materials analysis opportunities on the new neutron imaging facility IMAT@ISIS, J. Inst. 11 (2016)

[2] M.E. Fitzpatrick, M.T. Hutchings and P.J. Withers, Separation of macroscopic, elastic mismatch and thermal expansion misfit stresses in metal matrix composite quenched plates from neutron diffraction measurements, Acta Mater. 45 (1997) 4867–4876. https://doi.org/10.1016/S1359-6454(97)00209-7

[3] J.R. Santisteban, L. Edwards, A. Steuwer and P.J. Withers, Time-of-flight neutron transmission diffraction, J. Appl. Crystallogr. 34 (2001) 289–297. https://doi.org/10.1107/S0021889801003260

[4] A.S. Tremsin, J.B. McPhate, W. Kockelmann, J.V. Vallerga, O.H.W. Siegmund and W.B. Feller., Energy-Resolving neutron transmission radiography at the isis pulsed spallation source with a high-resolution neutron counting detector, IEEE Trans. Nucl. Sci. 56 (2009) 2931–2937. https://doi.org/10.1109/TNS.2009.2029690

[5] J.R. Santisteban, L. Edwards, M.E. Fitzpatrick, A. Steuwer, P.J. Withers, M.R. Daymond, M.W. Johnson, N. Rhodes and E.M. Schooneveld, Strain imaging by Bragg edge neutron transmission, Nucl. Instr. Meth. Phys. Res. A 481 (2002) 765–768. https://doi.org/10.1016/S0168-9002(01)01256-6

[6] A.S. Tremsin, T.Y. Yau and W. Kockelmann, Non-destructive Examination of Loads in Regular and Self-locking Spiralock ® Threads through Energy- resolved Neutron Imaging, Strain 52 (2016) 548–558. https://doi.org/10.1111/str.12201

[7] M.R. Daymond, M.W. Johnson and D.S. Sivia, Analysis of neutron diffraction strain measurement data from a round robin sample, J. Strain Anal. 37 (2002) 73–85. https://doi.org/10.1243/0309324021514844

MECA SENS 2017   Materials Research Forum LLC
Materials Research Proceedings **4** (2018) 155-160   doi: http://dx.doi.org/10.21741/9781945291678-24

Deformation Analysis of Reinforced Concrete using Neutron Imaging Technique

T. Koyama[1], K. Ueno[2], M. Sekine[1], Y. Matsumoto[3,a], T. Kai[4,b], T. Shinohara[4,c], H. Iikura[5,d], H. Suzuki[5,e*] and M. Kanematsu[2,f]

[1]Graduate School of Science and Technology, Tokyo University of Science, Noda, Chiba, Japan

[2]Faculty of Science and Technology, Tokyo University of Science, Noda, Chiba, Japan

[3]Research Center for Neutron Science and Technology, Comprehensive Research Organization for Science and Society, Tokai, Ibaraki, Japan

[4]J-PARC Center, Japan Atomic Energy Agency, Tokai, Ibaraki, Japan

[5]Materials Sciences Research Center, Japan Atomic Energy Agency, Tokai, Ibaraki, Japan

[a]y_matsumoto@cross.or.jp, [b]tetsuya.kai@j-parc.jp, [c]takenao.shinohara@j-parc.jp, [d]iikura.hiroshi@jaea.go.jp, [e]suzuki.hiroshi07@jaea.go.jp, [f]manabu@rs.noda.tus.ac.jp

Keywords: Neutron Imaging, Reinforced Concrete, Deformation Analysis, Image Analysis

Abstract. We suggest a novel method to observe internal deformation of concrete using a neutron transmission imaging technique. In order to visualize the internal deformation of concrete, cement paste markers containing Gd_2O_3 powder were dispersed two-dimensionally around the ferritic deformed rebar in reinforced concrete. Displacement of the neutron transmission image of the Gd marker was evaluated by a change in the position of the marker as a function of the travel distance of the vertical sample stage, and it was successfully evaluated to within approximately ±0.1 mm accuracy by analyzing selected markers with higher contrast and circularity. Furthermore, concrete deformation under pullout loading to the embedded rebar was evaluated in the same way and compressive deformation of concrete was successfully observed by analyzing the displacement of the markers. The results obtained in this study bring beneficial knowledge that the measurement accuracy of the marker displacement can be improved by choosing spherical-shaped markers and by increasing the contrast of markers.

Introduction

Reinforced concrete (RC), widely utilized for various architectural and civil engineering structures, is well known as a composite structure in which concrete with relatively low tensile strength and ductility is strengthened by reinforcements such as steel rods (rebars) with high tensile strength and ductility. The structural performance of RC is generally derived from the bond resistance between rebar and concrete. It has been demonstrated in previous studies that the neutron diffraction technique can be an alternative method to conventional strain gauges for evaluation of bond resistance by measuring the stress distribution of rebar embedded in concrete [1-3].

On the other hand, it is also important to evaluate the deformation behavior of concrete around the embedded rebar in order to discuss the mechanism of bond degradation between concrete and rebar for the RC structure. However, it is difficult to apply the neutron diffraction technique to the measurement of strain in concrete since high background noise scattering from hydrogen within the concrete makes diffraction measurements in concrete difficult. Alternatively, image analysis techniques such as a lattice method [4] and digital image correlation (DIC) [5], are commonly utilized for evaluating deformation of concrete quantitatively. They can assess concrete surface deformation by analyzing image contrast or marker displacement taken by a high resolution camera. In this study, therefore, we aim to develop a novel method to observe internal deformation of concrete using the neutron transmission imaging technique combined with a lattice method.

Experimental Procedure

Reinforced Concrete Specimen. Fig. 1 shows the schematic illustration of the RC specimen used in this study. A ferritic steel deformed-bar with a nominal 12.7 mm diameter to JIS3112 standard was embedded in a rectangular concrete specimen of size $50 \times 50 \times 130$ mm^3. The embedded length of rebar was 100 mm, and unbonded region of 30 mm in length was artificially introduced by surrounding the rebar with a polyvinyl chloride (PVC) pipe. In order to visualize the internal deformation of concrete around the embedded rebar, the cement paste markers containing 34 wt.% Gd$_2$O$_3$ powder (hereafter called "Gd marker") were dispersed two-dimensionally covering 15 % of the area of the A-A' cross section. The Gd markers were obtained by pulverizing a paste cylinder of $\phi 50 \times 100$ mm^3 having a specific Gd$_2$O$_3$ content and classifying the particle size into 0.6 to 1.7 mm. The RC specimen was demolded 72 hours after placing and then cured in water for 7 days. In order to reduce water in concrete that causes neutron attenuation by hydrogen, the RC specimen was placed in a constant temperature (20 ± 1 °C) and humidity (60 ± 5 RH %) room for 24 hours, and then was dried at 60 °C for 4 days before the neutron experiment.

Neutron Optical System. Fig. 2 shows the optical layout utilized in this study. The RC specimen mounted on the loading device, composed of a hydraulic jack and a load cell, was set up on a sample stage of BL22, RADEN [6], in the Materials and Life Science Experimental Facility (MLF) of the Japan Proton Accelerator Research Complex (J-PARC). RADEN is a next generation pulsed-neutron instrument, energy-resolved neutron imaging facility. The beam power of J-PARC MLF for this experiment was 150 kW, and the neutron flux at the sample position was estimated to be 5.0×10^5 n/cm^2/sec. The L/D ratio was set to 1000. The incident neutrons scattered and absorbed by the RC specimen were taken by a cooled CCD camera after being converted to visible light by a scintillator. The distance between the scintillator and the sample edge was set to 200 mm so as to reduce the influence of scattering from the specimen. A total of four images each with a 15 minute exposure time were taken with the same measurement configuration. The resolution of the transmission image was 2048×2048 pixels (16 bit), and the spatial resolution was approximately equivalent to be 0.2 mm. A displacement transducer with an accuracy of 0.01 mm was set up on the RC specimen to measure the displacement of the edge of the embedded rebar under pullout loading.

Experimental Condition. The analytical condition was, at first optimized by evaluating the displacement of the Gd marker images as a function of the travel distance of the sample stage in order to determine the position of the Gd marker on the transmission image accurately. The position of the Gd marker in a neutron transmission image of the RC specimen was evaluated by an image analysis technique using an open source image processing program,

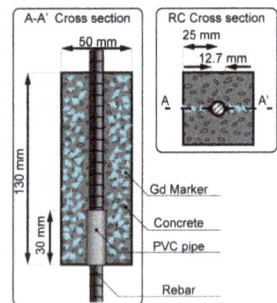

Fig. 1 Schematic illustration of the RC specimen used in this study.

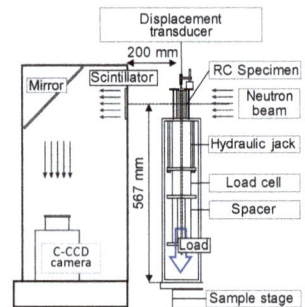

Fig. 2 Schematic optical layout for imaging experiment.

MECA SENS 2017   Materials Research Forum LLC
Materials Research Proceedings **4** (2018) 155-160   doi: http://dx.doi.org/10.21741/9781945291678-24

ImageJ 1.51n [7]. The transmission images were taken at 0.0, 0.1, 0.2, 0.3, 0.31, 0.33, 0.36 and 0.4 mm from the given original position by translating the sample vertically.

Secondly, the deformation of the RC specimen was measured under compressive deformation by analyzing the displacement of the Gd marker on the transmission image in order to verify the optimum condition of the image analysis determined by the above displacement measurement. The pullout loadings at 5, 10, 15, 20, 25 and 30 kN were applied to the rebar as shown in Fig. 2, making compressive deformation in the concrete part of the specimen. Pull out loading was manually controlled by the measured value of load cell installed in the loading device. The load on the rebar decreased drastically just after stopping the loading due to creep deformation. However, the displacement of the rebar measured by a displacement transducer was negligibly small, which means that concrete deformation around rebar was quite small during creep. Therefore, the transmission images started to be taken after waiting for 5 to 10 minutes after applying the load.

Results and Discussion

Transmission Image. Figure 3a shows a neutron transmission image of the RC specimen taken by RADEN. The Gd markers dispersed in the concrete matrix can be recognized in the image clearly. The dark area at the bottom of the center of the image represents the PVC pipe covering the rebar, as shown in Fig. 3a. In addition, the image contrast between concrete matrix and the embedded rebar can be observed as well.

Analysis of Gd Marker Position. The flow of the image analysis is shown in Fig. 3. The image analysis by ImageJ was conducted in an area of 940 × 1956 pixels as shown in Fig. 3a. At first, the transmission image was normalized by the incident neutron flux and the shade image (Fig. 3b). And then, the median filter was applied to some transmission images taken at the same sample position or at the same loading, in order to remove white spot noise and to smooth the image. After applied the median filter, a Fourier transform was applied and low frequencies in the transmission image were filtered by masking obtained by a 2D Fourier power spectrum less than 15 pixels in radius (Fig. 3c). After that, the spectrum obtained by excluding the low-frequencies was converted to the transmission image using an inverse Fourier transform procedure. This is known as a general high-pass filter, which can highlight the Gd marker images by reducing the background contrast variation representing the rebar and the PVC pipe. The transmission image obtained by the high-pass filter was binarized by defining an appropriate threshold (Fig. 3d). And then, using a morphological process, the spot noise on the obtained binary image was removed and the Gd marker images were shaped. Eventually, the Gd marker images were successfully extracted from the original transmission image as shown in Fig. 3e. After extracting the Gd marker images,

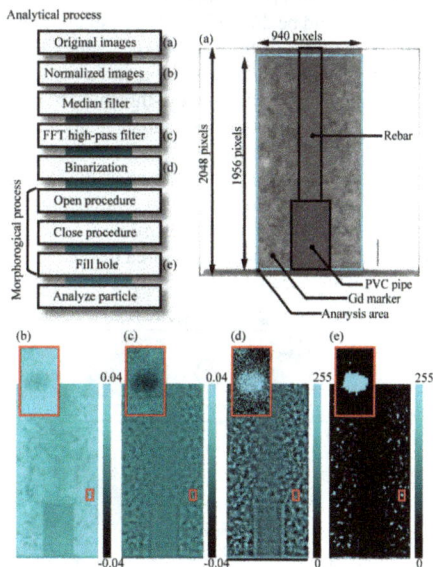

Fig. 3 Transmission image of the RC specimen (a), and the images for each process, (b), (c), (d) and (e), in the analytical procedure.

the center of mass (COM) for each marker was determined by analyzing particles using ImageJ. In the analysis, particles less than 300 pixel^2 in size (equivalent to the area of a circle with a diameter of 1 mm) were omitted to distinguish from the noise. Furthermore, the particles obviously different from the Gd markers were also regarded as noise and were manually omitted as well.

Evaluation of Marker Displacement. Displacement of the Gd marker image was evaluated by a change in the COM of the marker as a function of the travel distance of the vertical sample stage. As a result, the variation of the Gd marker displacement is quite scattered as shown in Fig. 4. The average displacement in Fig. 4b shows a linear variation in proportion to travel distance of the sample stage, with a standard deviation of ±4.5 pixels (equivalent to ±0.23 mm). This error might be due to the unstable shape of the Gd marker image. For instance, there is a case that the area, A_0, of the Gd marker image taken at an initial position of the sample stage is different from the area, A_i, of that taken at the next position of the sample stage (insertion individual marker images in Fig. 5a). This would be due to insufficient statistics compared with the contrast of the Gd marker image. The relation

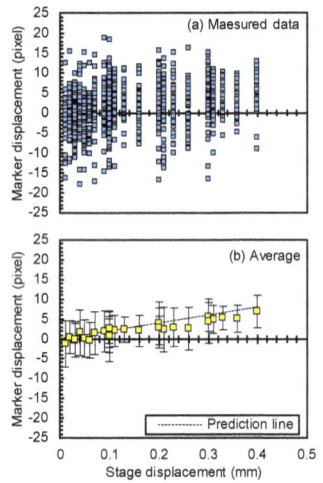

Fig. 4 The relation between marker and stage displacements. Prediction line in (b) corresponds to movement of the stage.

between the amounts of scattering and a threshold that gives an extraction limit of the Gd marker images were evaluated as a low-threshold to be defined by an area ratio, i.e. A_i/A_0 ($A_0>A_i$) or A_0/A_i ($A_i>A_0$). If choosing only Gd marker images that exceed a threshold, an increase in the area ratio, that is A_i approaches A_0, tends to decrease the scattering although the number of selectable Gd marker images decreases, as shown in Fig. 5a. Here, the standard deviation in Fig. 5a exhibits an average value of error bars taken at all sample displacements, and the number of the Gd marker images were counted in the transmission image taken at the initial stage position. Therefore, it was concluded that a change in the area of the Gd marker image, induced by an insufficient contrast, influences the accuracy of evaluation of the Gd marker displacement.

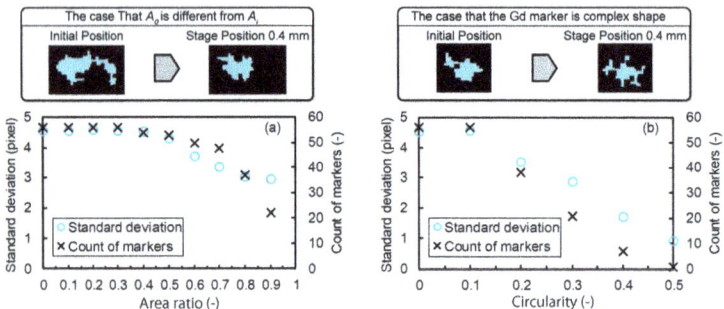

Fig. 5 The relations (a) between standard deviation and area ratio and (b) between standard deviation and circulality. Insertion images show examples of the change in the marker shape.

On the other hand, there is a case that Gd marker images with complex shape changes to a different shape at the next position of the sample stage (insertion individual marker images in Fig. 5b). Here, the relation between the standard deviation and the Gd marker images circularity, which were computed by ImageJ [7], was investigated. If choosing only Gd marker images above a given circularity, an increase in the circularity omits markers with a complex shape thereby decreasing the standard deviation as shown in Fig. 5b. Therefore, it was concluded that a change in the shape of the complex Gd marker image, induced by insufficient contrast, influences the accuracy of determination of the Gd marker displacement. Accordingly, the measurement accuracy of the displacement of the Gd marker might be improved by choosing the Gd marker images close to a circler shape.

Fig. 6 shows the Gd marker displacement evaluated at a low-threshold defined as an area ratio of 0.8 and a circularity of 0.3. At this threshold, the number of Gd markers decreased from 56 to 21. It can be found by comparing this result with Fig. 5 that the plots with large error were preferentially omitted by defining an appropriate threshold of area ratio and circularity. This result shows the scattering of ±2.2 pixels (equivalent to ±0.11 mm), which is approximately half of the result in Fig. 4. Therefore, the measurement accuracy of the Gd marker displacement can be improved by choosing a spherical shaped Gd marker, and by increasing the contrast of the Gd marker by increasing the amount of Gd_2O_3 powder contained in the marker and also by increasing statistical accuracy.

Deformation Analysis. Concrete deformation of the RC specimen was measured under pullout loading. The concrete part along the bonded region integrally shifts vertically toward the loading direction followed by the compressive deformation along the unbounded region. The transmission image was evaluated at a low-threshold defined as an area ratio of 0.8 and a circularity of 0.3. Applying this threshold, the number of acceptable markers at 5, 10, 15, 20, 25 and 30 kN is 3, 8, 22, 12, 0 and 6, respectively. As shown in Fig. 7, the Gd marker images move toward the loading direction proportionally to applied loading. Furthermore, the displacement of the Gd marker images are always evaluated to be slightly smaller than the displacement of rebar measured by a displacement transducer, while their trends agree well within the error bars. This small difference might be caused by asymmetric distortion of the RC specimen under pullout loading. Selecting the appropriate Gd marker with higher contrast and circularity enables us to analyze deformation of concrete accurately.

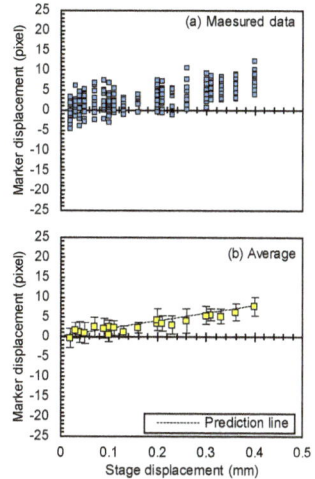

Fig. 6 The relation between marker and stage displacements obtained by defining threshold as an area ratio of 0.8 and a circularity of 0.3.

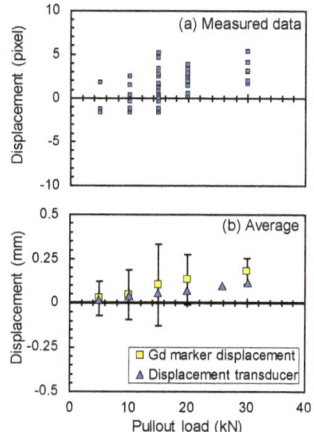

Fig. 7 The relation between marker displacement and applied loading, compared with the displacement measured by a displacement tranceducer.

MECA SENS 2017   Materials Research Forum LLC
Materials Research Proceedings 4 (2018) 155-160   doi: http://dx.doi.org/10.21741/9781945291678-24

Summary

We have suggested a novel method to observe internal deformation of concrete by the neutron transmission imaging technique with image analysis. In order to visualize the internal deformation of concrete around the embedded rebar, the displacement of the cement paste markers containing 34 wt.% Gd_2O_3 powder, dispersed two-dimensionally in the reinforced concrete samples were evaluated. Displacement of the Gd marker image was evaluated by a change in the position of the marker as a function of the travel distance of the vertical sample stage, and it was successfully evaluated within approximately ±0.1 mm accuracy by image analysis for selected markers with higher contrast and higher circularity. Furthermore, concrete deformation under pullout loading applied to the embedded rebar was evaluated by the same procedure and the compressive deformation of the concrete part was successfully observed by analyzing the displacement of the Gd marker images. The measurement accuracy of the Gd marker displacement can be improved by choosing a spherical shaped Gd marker, by increasing the contrast of the Gd marker and also by increasing statistics accuracy. This method is expected to be utilized for clarifying the bond mechanism between concrete and rebar by observing concrete deformation around rebar combined with the bond stress distribution measured by the neutron diffraction technique.

Acknowledgement

These measurements at J-PARC MLF were performed under a user program (Proposal No. 2016B0155). This work was supported by JSPS KAKENHI Grant Number 17K05125. We would like to acknowledge Mr. K. Satake of The University of Tokyo for his experimental assistance.

References

[1] H. Suzuki, K. Kusunoki, Y. Hatanaka, T. Mukai, A. Tasai, M. Kanematsu, K. Kabayama and S. Harjo, Measuring strain and stress distributions along rebar embedded in concrete using time-of-flight neutron diffraction, Meas. Sci. Technol. 25 (2014) 025602. https://doi.org/10.1088/0957-0233/25/2/025602

[2] H. Suzuki, K. Kusunoki, M. Kanematsu, A. Tasai, Y. Hatanaka, N. Tsuchiya, S.C. Bae, S. Shiroishi, S. Sakurai, T. Kawasaki and S. Harjo, Application of neutron stress measurement to reinforced concrete structure, JPS Conf. Proc. 8 (2015) 031006. https://doi.org/10.7566/JPSCP.8.031006

[3] H. Suzuki, K. Kusunoki, M. Kanematsu, T. Mukai and S. Harjo, Structural engineering studies on reinforced concrete structure using neutron diffraction, Mater. Res. Proc. 2 (2016) 25-30.

[4] K. Watanabe, H. Higashi, T. Miki and J. Niwa, Real Time Image Analyzing System for loading tests of structural concrete, Doboku Gakkai Ronbunshuu E 66 (2010) 94-106 [in Japanese]. https://doi.org/10.2208/jsceje.66.94

[5] K. Yu, J. Yu, Z. Lu and Q. Chen, Fracture properties of high-strength/high-performance concrete (HSC/HPC) exposed to high temperature, Mater. Struct. 49 (2016) 4517-4532. https://doi.org/10.1617/s11527-016-0804-x

[6] T. Shinohara, T. Kai, K. Oikawa, M. Segawa, M. Harada, T. Nakatani, M. Ooi, K. Aizawa, H. Sato, T. Kamiyama, H. Yokota, T. Sera, K. Mochiki and Y. Kiyanagi, Final design of the Energy-Resolved Neutron Imaging System "RADEN" at J-PARC, J. Phys. Conf. Ser. 746 (2016) 012007. https://doi.org/10.1088/1742-6596/746/1/012007

[7] C.A. Schneider, W.S. Rasband and K.W. Eliceiri, NIH Image to ImageJ: 25 years of image analysis, Nat. Methods 9 (2012) 671-675. https://doi.org/10.1038/nmeth.2089

Others

MECA SENS 2017   Materials Research Forum LLC
Materials Research Proceedings 4 (2018) 163-168   doi: http://dx.doi.org/10.21741/9781945291678-25

Influence of Hydrogenation on Residual Stresses in Oxygen-Implanted Ti-6Al-4V Alloy

S. Nsengiyumva[1,a*], T.P. Ntsoane[2,b], M. Topic[3,c], L. Pichon[4,d]

[1]Department of Physics and Electronics, Rhodes University, 6140 Grahamstown, South Africa

[2]Research and Development Division, Necsa Limited, P.O. Box 582, Pretoria 0001, South Africa

[3]iThemba LABS, National Research Foundation, P.O. Box 722, Somerset West 7129, South Africa

[4]Institute Pprime, UPR 3346CNRS, Université de Poitiers, ISAE-ENSMA, France

[a]s.nsengiyumva@ru.ac.za, [b]tshepo.ntsoane@necsa.co.za, [c]mtopic@tlabs.ac.za, [d]luc.pichon@univ-poitiers.fr

Keywords: Titanium Alloy, Oxygen Implantation, Hydrogenation, 2D XRD, Principal Stress

Abstract. We report the influence of hydrogenation on residual stresses in an oxygen-implanted Ti-6Al-4V alloy. Prior to hydrogenation, oxygen ions were implanted in Ti-6Al-4V samples at fluence 3×10^{17} ions/cm^2 with energies of 50 keV at room temperature and 550°C and 100 keV and 150 keV at 550°C. Hydrogenation was carried out on all samples at 550°C for two hours. Residual stresses were analysed by X-ray diffraction using the $\sin^2\psi$ method and components of in-plane principal stresses were determined. Our results show compressive stress relaxation in all samples implanted with 50, 100 and 150 keV at 550°C as compared to unimplanted sample. Subsequent to hydrogenation, a stress shift to tensile side is observed in all implanted samples at 550°C.

Introduction

Owing to its high strength to weight ratio and good corrosion behaviour, Ti-6Al-4V is by far the most common Ti alloy used in a broad range of aerospace, marine, industrial, medical and commercial applications [1-3].

In recent years, Ti-6Al-4V has also attracted increasing research interest as a potential candidate for hydrogen storage due to its high affinity to absorb and release hydrogen [4, 5]. Moreover recent studies [6, 7] have suggested that a titanium oxide layer could enhance hydrogen absorption and promote Ti-6Al-4V alloy as an efficient hydrogen storage material. It has been established that the rate of hydrogen diffusion is higher by several orders of magnitude in the β phase than in the α phase and that fully lamellar microstructure (β phase) would absorb more hydrogen than duplex microstructure [8]. Furthermore the higher solubility, as well as the rapid diffusion of hydrogen in the beta titanium results from the relatively open body centered cubic structure, which consists of 12 tetrahedral and   octahedral interstices in comparison to 4 tetrahedral and 2 octahedral interstitial sites in the hexagonal closed packed lattice of alpha titanium [6]. The presence of hydrogen in both α and β phases results in lattice expansion and the associated strain/stress fields could give rise to altered physical properties [9]. The aim of this study is to investigate the effect of hydrogenation on residual stresses of oxygen-implanted Ti-6Al-4V alloy to obtain a better understanding on the relationship between the hydrogen absorption and the subsequent induced stress. It is worth mentioning that the amount of hydrogen absorbed in the samples under investigation has been reported in our recent publication [10].

Experimental
The material used in this study was Ti-6Al-4V alloy (grade-5) with typical composition of 6 wt.% Al, 4 wt.% V, 0.25 wt.% Fe_{max}, 0.2 wt.% O_{max} and Ti as reminder (corresponding approximately to the atomic composition Ti0,86Al0,10V0,04). The alloy was supplied in a form of annealed rod by Goodfellow©. The samples used were cut into disks (10 mm diameter and 2 mm thick), mounted in resin and polished using the following procedure. They were first grinded using 800 grit and 1200 grit SiC papers; after each grinding step, the samples were rinsed with water and cleaned ultrasonically in ethanol. Further polishing was performed using diamond paste with 9 μm grain size. After cleaning, a mixture of OP-S (colloidal silica suspension) and H_2O_2 in 5:1 ratio was used for final polishing of the samples.

Oxygen O^+ ions at a fluence of 3.0×10^{17} ions/cm^2 were implanted with 50, 100 and 150 keV energy and at a beam current of 10 μA using in-line implanter Eaton NV 3206. The implantation was performed at room temperature and 550°C.

All the samples were afterwards subjected to hydrogenation at 550°C for 2 hrs in 15 % hydrogen and 85 % argon mixture at 1 atm pressure. The gas flow was kept constant at 14 cm^3/s. The samples were heated up to hydrogenation temperature and cooled down to room temperature at a rate of 5°/min under vacuum of 5.5 x 10^{-2} mbar. Based on our previous experimental work, the optimum temperature for hydrogen absorption in the Ti-6Al-V alloy was determined as 550 °C [10, 11].

X-ray stress investigation was done using a Bruker D8 Discover diffractometer equipped with a ¼ Eulerian cradle. The primary side optics included a graphite monochromator and a 0.8 mm collimator. No secondary optics on the detector side. Measurements were performed using a copper tube operating in spot focus mode, employing the ψ-tilt method. For a full stress tensor determination, measurements were done at six azimuth orientations, $\phi = 0°$, 45°and 90°, and ϕ +180 = 180°, 225° and 270° with each azimuth measured at eight tilt angles $\psi = 0°$, 10°, 20°, 30°, 40°, 50°,60° and 70°. The rotation of ϕ by 180° allowed measurement at negative tilt angles. Diffracted data was collected using a 2-D Vântec 500 detector. The diffraction peak corresponding to (211) at d = 0.93217 Å was used to measure residual strain. It was observed that some degree of texture was present in the samples and it was more pronounced in the hydrogenated samples (Fig. 1a) than in implanted samples (Fig. 1b). To minimise the texture effects on the analysis, samples were oscillated in the X-Y plane during measurements. The stress analysis of the data was done using the manufacturer proprietary software, Leptos v6.2. The maximum penetration depth of X-rays was found to be about 6 μm.

MECA SENS 2017   Materials Research Forum LLC
Materials Research Proceedings **4** (2018) 163-168   doi: http://dx.doi.org/10.21741/9781945291678-25

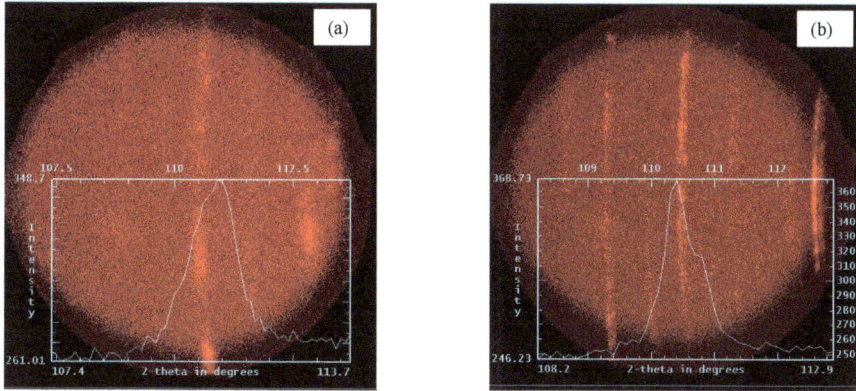

Fig. 1: 2D diffraction image of the reflection (211) used for strain measurement. (a) highly textured hydrogenated Ti-6Al-4V, (b) The texture is less pronounced in implanted Ti-6Al-4V alloy.

Theory of 2D XRD stress determination

Stress measurement with two-dimensional X-ray diffraction (XRD^2) is based on the fundamental relationship between the stress tensor and the diffraction cone distortion. The diffraction peak $2\theta_0$ shifts are measured along the diffraction rings. Since a diffraction ring in a 2D pattern contains far more data points than a conventional diffraction peak, an XRD^2 system can measure stress with higher accuracy and requires less data collection time, especially in dealing with highly textured materials, large grain size, small sample areas, weak diffraction, stress mapping, and stress tensor measurement [12].

The strain tensor ε_{ij}, determined in the sample coordinate system, and the distortion of the diffraction ring in a particular direction ($\eta,2\theta$) are related by the two-dimensional fundamental equation for strain measurement

$$f_{11}\varepsilon_{11} + f_{12}\varepsilon_{12} + f_{22}\varepsilon_{22} + f_{13}\varepsilon_{13} + f_{23}\varepsilon_{23} + f_{33}\varepsilon_{33} = \ln\left(\frac{\sin\theta_0}{\sin\theta}\right),$$ (1)

where f_{ij} are the strain coefficients, $\ln\left(\frac{\sin\theta_0}{\sin\theta}\right)$ is the diffraction cone distortion for a particular ($\eta,2\theta$) position, θ_0 and θ are the peak position for the stress-free material and stressed material, respectively.

In general, the stress tensor can be calculated from the measured strain tensor by Hooke's law. The stress tensor can be expressed in linear form as

$$p_{11}\sigma_{11} + p_{12}\sigma_{12} + p_{22}\sigma_{22} + p_{13}\sigma_{13} + p_{23}\sigma_{23} + p_{33}\sigma_{33} = \ln\left(\frac{\sin\theta_0}{\sin\theta}\right),$$ (2)

where p_{ij} are stress coefficients.

MECA SENS 2017   Materials Research Forum LLC
Materials Research Proceedings 4 (2018) 163-168   doi: http://dx.doi.org/10.21741/9781945291678-25

For this investigation the true stress-free theta value θ_0 was estimated from measurement of the side opposite to the implantation side. Any error in stress-free theta (stress-free lattice spacing) contributes only to pseudo-hydrostatic term [12] as $\sin^2\psi$-method relies on the slope of plot and not the intercept.

Results and discussion

Figure 2 shows in-plane principal stresses σ_1 and σ_2 for unimplanted and implanted samples as a function of implantation energies of 50 keV (at room temperature and 550°C), 100 keV and 150 keV at 550°C. In Fig. 2a the principal stress component σ_1 of the unimplanted sample is compressive with a value of about 350 MPa. Upon implantation at 550°C with 50 keV, the stress value decreases to about 190 MPa. With further increased implantation energy, the stress value decreases to 298 MPa and 228 MPa for 100 keV and 150 keV respectively. The decrease in the stress value for samples at 550°C points to stress relaxation brought about by ion implantation. It can be observed however, that, at room temperature implantation with 50 keV oxygen ions, the stress component σ_1 does not undergo stress relaxation; instead, the stress becomes more compressive, shifting from 340 to 410 MPa. In the case of the principal stress σ_2 (Fig. 2b), the stress relaxation is observed in samples implanted at room temperature as well as in samples implanted at 550°C. The unimplanted sample has a stress value of 322 MPa. With implantation energy of 50 keV at room temperature, the stress relaxes to 290 MPa whereas it relaxes to 165 MPa, 120 MPa, and 98 MPa for samples implanted at 550°C for 50 keV, 100 keV and 150 keV respectively.

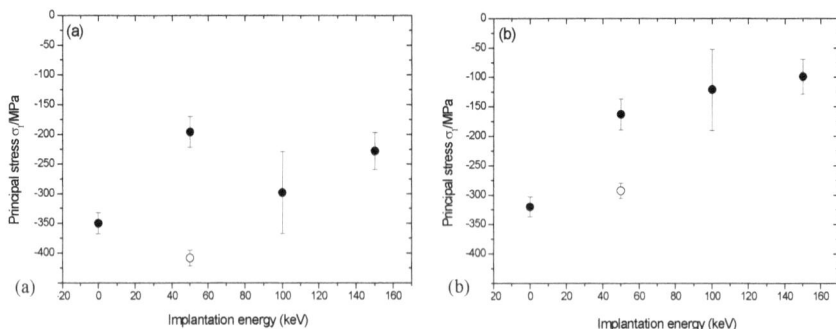

Fig. 2: Principal stress as a function of implantation energy for σ_1 component (a) and σ_2 component (b) for unimplanted sample, implanted samples with 50 keV at RT (open circle) and 50 keV, 100 keV and 150 keV at 550°C (filled circles).

Figure 3 shows in-plane principal stresses for hydrogenated unimplanted sample and hydrogenated implanted samples as a function of implantation energies: 50 keV at room temperature and 50 keV and 150 keV at 550°C. The hydrogenation of the sample implanted with 100 keV is not shown in Fig. 3. The results obtained from this sample were deemed unreliable. Subsequent to hydrogenation at 550°C for two hours, it can be observed from Figs. 3a and b that the components σ_1 and σ_2 that were compressive upon implantation have become tensile. Only the stress component σ_2 of the unimplanted sample remains compressive following hydrogenation.

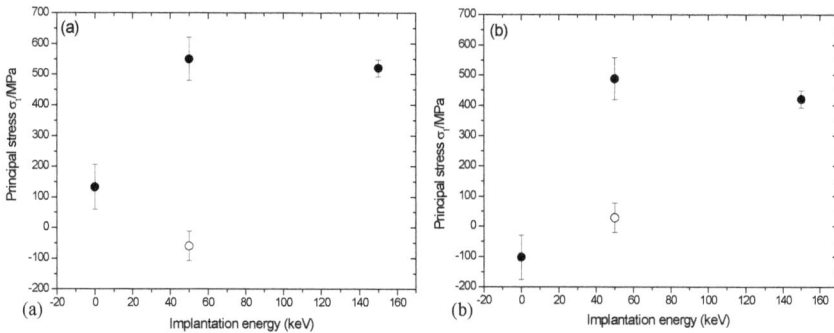

Fig. 3: Principal stress as a function of implantation energy after hydrogenation for component (a) and component (b) for unimplanted sample, implanted samples with 50 keV and 150 keV.

The effect of hydrogenation on both stress components σ_1 and σ_2 with respect to unimplanted and implanted samples is shown in Fig. 4. In the case of σ_1 (Fig. 4a), the stress value changes by about 500 MPa for the unimplanted sample, 750 MPa and 725 MPa for implanted samples with 50 keV and 150 keV respectively. In the case of σ_2 (Fig. 4b), the stress value changes by 218 MPa for the unimplanted sample, 650 MPa and 520 MPa for implanted samples with 50 keV and 150 keV respectively.

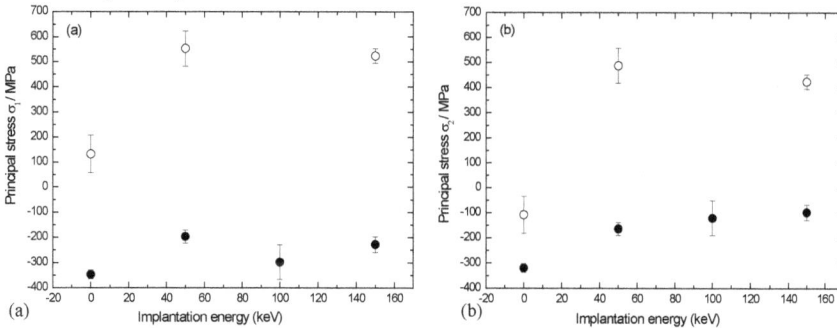

Fig. 4: Principal stress as a function of implantation energy showing the change of stress following hydrogenation. (a) σ_1 stress component. (b) σ_2 stress component. The full circles correspond to the stress values of unimplanted and implanted samples and the open circles correspond to the stress values of the same samples but after hydrogenation.

Conclusion
Residual stresses induced by ion implantation and hydrogenation in Ti-6Al-4V alloy have been investigated. Samples analysed were implanted with oxygen ions at 3×10^{17} ions/cm^2 with energies of 50 keV at room temperature and 550^{0}C, 100 keV and 150 keV at 550^{0}C. Hydrogenation was carried out at 550^{0}C for two hours. It was observed that oxygen implantation in Ti-6Al-4V alloy brings about stress relaxation of the pre-existing compressive stresses. Subsequent to hydrogenation, a stress shift from compressive to tensile was observed in all implanted samples at 550^{0}C.

Acknowledgements
We are grateful to Rhodes University for financial support. We would also like to thank Marc Marteau at the Pprime Institute (France) for his assistance in ion implantation.

References

[1] M. Peters, J. Kumpfert, C.H. Ward and C. Leyens, TitaniumAlloys for Aerospace Applications, in: Titanium and Titanium Alloys, Adv. Eng. Mater. 5 (2003) 419-427. https://doi.org/10.1002/adem.200310095

[2] I. Gurrapa, Characterization of titanium alloy Ti-6al-4V for chemical, marine and industrial applications, Mater. Charact. 51 (2003) 131-139. https://doi.org/10.1016/j.matchar.2003.10.006

[3] C.N. Elias, J.H.V. Lima, R. Valiev and M.A. Meyers, Biomedical application of titanium and its alloys, JOM 60 (2008) 46-49. https://doi.org/10.1007/s11837-008-0031-1

[4] L. Bonaccorsi, L. Calabrese, A. Pintaudi, E. Proverbio, F. Aliotta, R. Ponterio , A. Scherillo and D. Tresoldi, Reversible hydrogen absorption in a Ti-6Al-4V alloy produced by mechanical alloying, Int. J. Hydrog. Energy 39(28) (2014) 15540-15548. https://doi.org/10.1016/j.ijhydene.2014.07.149

[5] A. López-Súarez, J. Rickards and R.Trejo-Luna, Mechanical and microstructural changes of Ti and Ti-6Al-4V alloy induced by the absorption and desorption of hydrogenJ. Alloys Compd. 457 (2008) 216-220. https://doi.org/10.1016/j.jallcom.2007.03.031

[6] A. López-Súarez, Influence of surface roughness on consecutively hydrogen absorption cycles in Ti–6Al–4V alloy, Int. J. Hydrog. Energy 35 (2010) 10404-10411. https://doi.org/10.1016/j.ijhydene.2010.07.163

[7] J.L. Blackburn, P.A. Parilla, T. Gennett, K.E. Hurst, A.C. Dillon and M.J. Heber, Measurement of the reversible hydrogen storage capacity of milligramTi–6Al–4V alloy samples with temperature programmed desorption and volumetric techniques, J. Alloys Compd. 454 (2008) 483-490.

[8] H.G. Nelson, D.P. William and J.E. Stein, Environmental hydrogen embrittlement of an α-β titanium alloy: Effect of microstructure, Metall. Mater. Trans. B 3 (1972) 473-479. https://doi.org/10.1007/BF02642051

[9] L. Miaoquan, Z. Weifu, Z. Tangkui, H. Hongliang and L. Zhigiang, Effect of microstructure of Ti-6Al-4V alloys, Rare Metal Mat. Eng. 39(1) (2010) 1-5. https://doi.org/10.1016/S1875-5372(10)60071-9

[10]M. Topic, L. Pichon, S. Nsengiyumva, G. Favaro, M. Dubuisson, S. Halindintwali, S.Mazwi, J. Sibanyoni, C. Mtshali and K. Corin, The effect of surface oxidation on hydrogen absorption in Ti-6Al-4V alloy studied by elastic recoil detection analysis (ERDA) and nanohardness techniques, J. Alloys Compd. 740 (2018) 879-886. https://doi.org/10.1016/j.jallcom.2017.11.269

[11]S. Mazwi, Hydrogen storage in Ti-based coatings and Ti6Al4V alloy, Unpublished, Master's thesis, University of the Western Cape, Cape Town, South Africa.

[12]B. He Bob, Two-Dimensional X-Ray Diffraction, Wiley, 2009.

Keyword Index

Author Index

About the Editors

The editors comprises three generations of researchers all with a passion for understanding and describing the physical world.

Deon Marais holds a B.Eng in computer and electronics and an M.Eng and Ph.D. in nuclear engineering from the North-West University, South Africa. In 2009 he started his career as a nuclear software engineer on high temperature gas cooled reactor projects whereafter he moved to Necsa (South African Nuclear Energy Corporation) SOC Ltd where he became responsible for the data acquisition & control, reduction and visualisation systems of the neutron diffraction facility. He is also an instrument scientist on the MPISI neutron strain scanner.

Thomas M. (Tom) Holden received his B.Sc. and Ph.D. from Leeds University in England. He began working in the field of neutron scattering during post-doc positions at the United Kingdom Atomic Energy Authority, Harwell and at Chalk River Nuclear Laboratories in Canada. In 1966 he joined the staff of Atomic Energy of Canada (AECL) and studied magnetism until the 1980's. In 1983 he began working on measurements of residual stress and texture in engineering components and materials by neutron diffraction. In 1998 he retired from AECL and since then has consulted in Australia, the Netherlands, Japan, Korea, the United States and South Africa.

Andrew M. Venter received his qualifications up to the level of Ph.D. from the Rand Afrikaans University (now University of Johannesburg) and has an an extraordinary professorship with the North-West University in South Africa. He is a scientific group leader at Necsa and an National Research Foundation of South Africa rated research scientist with more than 30 years' experience in neutron and X-ray diffraction techniques. He has been responsible for the establishment of world-class neutron and X-ray diffraction facilities at Necsa in support of material science studies in South Africa.

Dr Deon Marais
Dr Tom Holden
Prof Andrew Venter

www.ingramcontent.com/pod-product-compliance
Lightning Source LLC
Chambersburg PA
CBHW071228210326
41597CB00016B/1988